Qualitative Modeling of Complex Systems

Qualitative Modeling of Complex Systems
AN INTRODUCTION TO LOOP ANALYSIS AND TIME AVERAGING

Charles J. Puccia and Richard Levins

Harvard University Press
Cambridge, Massachusetts, and London, England   1985

Copyright © 1985 by Charles J. Puccia and Richard Levins
All rights reserved
Printed in the United States of America
10 9 8 7 6 5 4 3 2 1

This book is printed on acid-free paper, and its binding materials have been chosen for strength and durability.

LIBRARY OF CONGRESS CATALOGING IN PUBLICATION DATA

Puccia, Charles J.
  Qualitative modeling of complex systems.

  Bibliography: p.
  Includes index.
  1. System analysis.   I. Levins, Richard.
II. Title.   III. Title: Loop analysis.
QA402.P82   1985   003   85-5537
ISBN 0-674-74110-2 (alk. paper)

Preface

Both the internal development within the biological and social sciences and their application to problems of human urgency have confronted us with systems of increasing complexity for which neither our formal training nor our intuition is adequately prepared. We have to deal with a large number of variables, often belonging to different disciplines, interacting indirectly as well as directly. Often we lack precise information about the form of these interactions. We have few guidelines for deciding which simplifications of the problem are legitimate, that is, which simplifications enhance intelligibility without sacrificing fundamental properties.

The qualitative methods developed in this volume were designed to assist in understanding the behavior of such partly specified systems, to guide experimental design and data interpretation, and to supplement the more familiar large-scale quantitative methods made possible by improved computer technology. Both approaches presented here—loop analysis and time averaging—were developed in the 1970s and have been used mostly in theoretical and applied ecology. They have not been adopted widely, however, because no simple exposition of the procedures and the approach to complex systems analysis has been available. We hereby offer that exposition.

This book will enable students, researchers, and field workers in academia, the private sector, and governmental agencies to analyze systems of interest to them by using qualitative methods. Moreover, it will help to reorient their thinking about how to face complex systems. It will help them to recognize the complexity of the systems, yet not be overwhelmed by it.

There has been a slight attempt to keep the mathematics at the same level throughout—with the exception of the last two chapters, which clearly are more advanced. Although the material may be unfamiliar at times, it presumes only an intermediate level of mathematics and therefore will be comprehensible to most.

The order of the chapters is somewhat arbitrary, although we wrote them with the idea that most students would start with the first chapter and work toward the last. Chapter 1 presents the motivation and philosophical rationale behind modeling, and in particular the use of qualitative models. The next two chapters (2 and 3) develop the algebraic algorithms of loop models. These easily could follow Chapters 4 and 5, allowing for the results simply to be taken on faith, to see the uses of qualitative modeling and how to transfer an idea about the way some part of the world works into a model.

Chapter 6 is entirely optional. It is included for those who feel dissatisfied with being given an algorithm without support for why it works; it stops short of rederiving dynamical systems theory.

Slightly more advanced mathematics is required for Chapter 7, on time averaging. Yet it is self-contained, so most of the material should be understandable to anyone with a little knowledge of differential equations and the statistical operators of average, variance, and covariance. The flourish of notation is merely a camouflage for straightforward algebraic operations.

All the chapters include examples that illustrate or extend the material presented up to that point in the text. They are written so the reader may first attempt to solve the problem posed and then consider its fairly detailed solution, which is given in the hope that this volume can be a practical workbook as well as a text.

We thank Michel Loreau for his painstaking efforts in reading the entire manuscript and for his special attention to correcting errors in the mathematics. Janice Alpert, Pamela Anderson, David Glaser, and Peter Taylor provided us with detailed criticism of the first five chapters. Many of their suggestions have been incorporated into the book and have greatly improved its usefulness. We are indebted to them for their assistance.

We wish to acknowledge the publishers' kind permission to reproduce the following: Figure 4.9 from *Geology*, 9:491, copyright © 1981, Rudy Slingerland; Figure 5.10 and Table 5.1 from *Nature*, 273:288, copyright © 1978, Macmillan Journals Limited; Figure 5.11 from *The Population Ecology of Cycles in Small Mammals* by James P. Finerty, p. 184, copyright © 1980, Yale University Press; Figures 5.14 and 5.15 from the *Transactions of the American Fisheries Society*, 110:778 and 780, respectively, copyright © 1981, American Fisheries Society.

C.J.P.
R.L.

Contents

1 **Qualitative Models** 1
   Uses of Models 2
   Qualitative versus Quantitative Models 4
   Preview of Models 6
   Limitations of Models 7
   Strategy for Qualitative Model Applications 10

2 **Loop Models: Definition of Terms and Calculation of Model Properties** 12
   Symbols of Signed Digraphs 12
   Notation 14
   Paths, Links, and Loops 14
   Conjunct and Disjunct Loops 17
   Feedback 17
   Stability 24

3 **Loop Models: Predicting Change** 32
   Parameter Change 32
   Open Paths and Complementary Subsystems 36
   Changes in Equilibrium Abundance 40
   Tables of Predictions 45
   Links from Equations 51

4 **Qualitative Predictions** 53
   Equilibrium Levels and Turnover Rates 54
   Correlations from Predictions 55
   From Model to Predictions 59
   Some General Observations 74
   Lumping Variables 79
   Model Ambiguity 84
   Experimental Design and Loop Models 90
   Correlation Patterns 95

5 From Concept to Model 119

Ecological Examples 127
Changes in the Graph Structure 137
From Data to Graph 139

6 Dynamical Systems and Loop Analysis 150

Equilibrium 150
Stability 152
Self-Effect Terms (Self-Damping) 159
Signed Digraphs, the Characteristic Equation, and Feedback 162
Eigenvectors 173
Discrete Systems 190

7 Time Averaging 194

Definitions 195
Operations with Time Averaging 199
Finite Intervals 216
Discrete-Time Processes 223
Time Averaging and Loop Analysis 224
Linear Equations 225
Nonlinear Equations 226

Appendix: Some Mathematical Operations and Concepts 237

References 253

Index 255

Qualitative Modeling of Complex Systems

1 Qualitative Models

We often are told that modern life is increasingly complex. Several different ideas are included within this notion of complexity. First, the number of different kinds of things has grown enormously. In the United States we manufacture more than fifty thousand different synthetic chemicals, many of which become—even if briefly—a part of our environment. Many others then are derived from these by natural processes in the atmosphere, soil, or living bodies. In addition to the great diversity of physical and chemical objects, we have created social and economic entities —laws, indicators, policies, statutes, and beliefs—which although themselves have no physical existence, nevertheless influence the appearance, movement, and disappearance of natural and synthetic substances.

There is also an increased complexity of interaction. As our own activities penetrate more deeply into the natural world, their impacts spread farther and last longer, so that distant events meet and interact in often unexpected ways.

In other cases the complexities always have been there, but are increasingly forced to our attention by the internal developments within the sciences or the demands that society makes of the applied sciences. Whereas previously the disciplines of evolution, ecology, biogeography, and population genetics were quite content to study different objects, we now see species populations as heterogeneous in age distribution, physiological states, and genetic makeup, with all of these varying in space and time through interactions with other equally heterogeneous and dynamic populations and physical conditions.

The epidemiology of infectious disease once was treated as a problem of the movements of microbes. Now we must examine the nutritional condition of people as it affects their susceptibility to disease; how their movement patterns affect exposure; and what their economic and social resources are for responding to illness. Thus, epidemiology must include physiology, microbiology, and sociology.

As the sciences have developed, we have been most successful in the detailed characterization of pieces of problems. But we have been ill prepared for solving, or even posing, broad problems. Most of the major failings of contemporary applied sciences have come about despite an increasing precision in the small, out of failure to recognize the consequences of our actions as they percolate through networks of causes that feed back on themselves, enhancing some processes and buffering others. Infant formulas that reduce breast feeding, agricultural schemes that increase malaria, clinical programs that ignore nutrition, industrial innovations that poison the workers, plant breedings that squeeze out genetic diversity, and water management projects that improve irrigation but reduce fish catch all illustrate the present difficulties—both intellectual and administrative—in examining complex systems. This is not surprising. The training we receive in "scientific method" emphasizes how to isolate and simplify problems, how to vary effects one at a time, and how to amass all the relevant "facts" before theorizing. Growing specialization exacerbates the problem, making us believe that the world is structured according to the table of organization of our own institutions. The final result is a growing sophistication in studying the small and an increasing irrationality in handling the whole.

Uses of Models

A model is an intellectual construct we study instead of studying the world. Every model distorts the system under study in order to simplify it. The world we wish to understand—ecological, social, economic, political—requires simplification to help us achieve an understanding of its systems.

There are two dangers in model building: one is that the model does not tell us about the world; the other is that it is a faithful representation, and therefore we are overwhelmed (see Fig. 1.1). Simplification is both legitimate and necessary as long as we are cautious, are willing to change the original underlying assumptions as necessary and build new models, and carefully interpret the model's predictions.

Legitimate simplification depends not only on the reality to be described but also on the state of the science involved. For example, the state of science in the early pioneering work in population genetics of the 1930s carried out by Fisher, Haldane, and Wright brought forward models that assumed the environment is constant. Each of these authors certainly was aware that this is not so. Yet the state of the science required that the following question be answered: could weak natural selection account for evolutionary change? For this problem it would have been unnecessarily complicated to worry about a changing environment. Today, of course, the questions asked in population genetics are different, and environmental heterogeneity is an essential ingredient of some models.

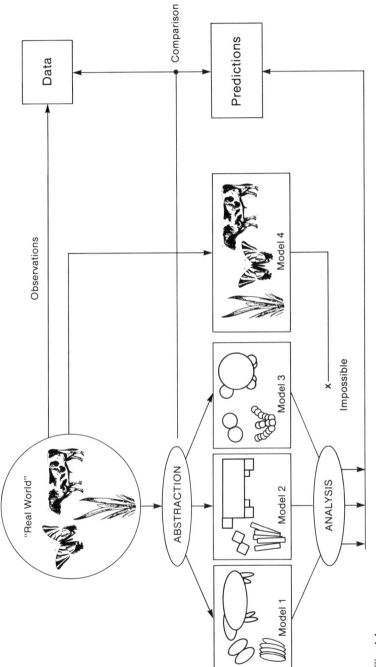

Fig. 1.1

Qualitative versus Quantitative Models

In a scientific world, where the demand for ever-greater precision seems a logical consequence of past successes, the expectation is for modeling methodologies that incorporate the capabilities of precise measurements. Models should be developed to make predictions or to explain properties of systems with equally precise quantitative statements. Nonetheless, we present herewith a modeling method where precision is not of interest and quantification is not of prime importance.

Several approaches to the study of complexity have been taken. The reductionist pathway asserts that the complexity we perceive is only apparent, that if we reduce the problem properly we will find the molecular basis, or the unit of pairwise species interaction, or the atoms of economic behavior which then can be assembled, at least in principle, to give us the whole system.

A second approach makes use of the vast capacity of computers to store information and to perform rapid computations. Much of systems analysis depends on simulation techniques—the representation of many variables within the computer, the careful measurement of parameters, the proposal of precise equations for the interactions among variables, and then, by numerical methods, the prediction of the outcome of the process. These methods are most successful where the parts of the system have been produced in isolation from one another, their behaviors precisely specified, and the circuit diagrams of their patterning carefully designed. Thus the simulation of engineering systems has shown very strong predictive power.

For systems whose parts are not listed in a catalog, which evolve together, which are difficult to measure, and which show unexpected capacity to form new connections, the results of simulation techniques have been less impressive. The masses of data required make the procedure very costly; the demand for quantitative precision often forces the exclusion of many variables, especially those involving human inputs; the predictions apply only to the original single system from which the models are derived, and are not easily extended even to similar objects; and the output often may be masses of numbers which allow some prediction but little understanding.

Our own approach to complex systems stresses qualitative understanding as the primary goal, rather than numerical prediction. Instead of accumulating all the relevant information, we see how much we can avoid measuring and still understand the system. We use qualitative analysis to determine which measurements are really necessary. Sometimes the qualitative analysis is sufficient for prediction or to guide experimentation. At other times it is a preliminary overview of a problem that permits a more systematic use of simulation methods.

The quantitative approaches to complex problems have had their greatest success in three general areas. The first is engineering design problems, where there is a single goal, although often opposing constraints. Precise quantitative results are achievable and the design possible, because the components are known. Second, in the areas of control and optimization, such as guidance systems or information processing, the quantitative approach also works well. As in problems of engineering design, often there are multiple constraints but a common goal. A major feature of this area of application is the ability to locate sensors and, in nearly unrestricted numbers within the components of the system, to measure the variables or parameters of interest. The third area of successful quantitative approach is in problems where reductionism or a narrowing of the boundaries of inquiry can be achieved. Here, multiple constraints get reduced. A model for understanding neurological activity, for example, becomes a model of the firing of a single axon under specified calcium and potassium concentrations.

Conversely, we are studying complex ecological and social problems that cannot be modeled in the ways just described, for these reasons:

1. Nature cannot be controlled, in the sense of creating uniformity.

2. Conflicting interests are always present, either among the organisms in nature or the goals of those wishing to intervene.

3. There are real variables that are either nonquantifiable or change in value with any attempt to measure them. For example, stress is a real variable in a diabetes model but any attempt to measure stress is itself stress inducing.

4. In a complex system of only a modest number of variables and interconnections, any attempt to describe the system completely and measure the magnitude of all the links would be the work of many people over a lifetime.

5. The question of interest may be not a particular system (for example, a particular lake or forest) but a whole class of systems with some similarity of structure. We may want to look at lakes in general, or at networks such as a food web organized into trophic levels with interactions only between adjacent levels, or at chemical networks in which the end points of a reaction inhibit the enzyme that catalyzes that reaction. Numerical results are inadequate for understanding what makes different systems behave similarly and form some sort of class.

6. In biological problems the search for quantification of links or establishment of accurately measured baselines ignores the fact that often the biological reality resides in the rules of construction of the system and not in the absolute values. The general character of a lake

district is not that each lake's ecosystem has exactly the same parameter values as the next, but that each may follow the same rules of trophic and biotic construction.

In the domain of ecological and social problems the qualitative approach has several advantages over the quantitative: It allows us to include variables which are difficult or even impossible to measure. A model of diabetes might include glucose, insulin, and other chemicals, but also anxiety. An environmental analysis might include the reproduction of fish, the rate of fishing, and a coefficient of bureaucratic inertia. As long as we can know the direction of effect of one variable on another, we do not require precise measurement or even the mathematical form of the effect.

Qualitative methods make it easier to include regulators or the experimentalists within the system, as part of the model rather than as external givens. Moreover, qualitative methods are inexpensive. When in doubt about the actual structure of a system, we can model alternatives and find out which differences matter. Therefore the process of qualitative modeling can be public, reproducible, and intelligible in a way that a large simulation project cannot be.

The conclusions derived from qualitative analysis do not allow for precise prediction or decisions about optimal courses of action, but they do permit us to decide the particular directions to push on a system in order to move it in a desired direction. Often this is all we can expect to do with living and, especially, human systems, which generate their own spontaneous activities.

Preview of Models

In the next three chapters we introduce models for the purpose of explaining qualitative models and analysis. To simplify, we omit proofs or justifications for the statements we make. In the later chapters we will discover that some statements are applicable within certain bounds, that they may have exceptions, and/or that they do not include possible alternatives. In short, we ask the reader to suspend judgment in the first three chapters to become familiar with loop model analysis.

Models are chosen to represent a diverse array of problems in marine, terrestrial, and applied ecology; in population ecology; and in social epidemiology. There is no need to be an expert in these disciplines to understand the models or their analyses. We make no attempt to extend the scope of knowledge within any subject, but use existing knowledge to introduce qualitative modeling methods. Moreover, purely hypothetical constructs will be used whenever they aid the explanation of qualitative analysis.

Loop models typically are presented by first describing the interactions between the objects of interest—the variables—before proceeding with the analysis of the model. Learning to make these interconnections is one part of building models and is explained in this book. Learning to recognize how interactions between and among variables produce the interconnections shown in the models is learning to translate an idea about the world into a loop model. The derivation of the initial concepts of the world, however, falls outside the scope of the work presented here.

One final word needs to be said about the procedure of translating ideas into models. Abstraction is the crucial step in model building. In the case of qualitative models it may be even more important than it is for quantitative models. In qualitative versus quantitative models the emphasis shifts from precisely measuring parameters to recognizing as many of the relevant variables and ways of associating or grouping them as one can. This means going beyond the knowledge of a single academic discipline. Furthermore, it means asking questions about the subject in a different way. For example, "Foxes eat hares" is a predator-prey relation. Questions about this relationship can be asked at the individual level, such as, What is the influence of a fox on a hare? But the question also can be raised to the population level: Are foxes important to hare populations? A relation at one level does not mean it has the *same* relation or the *same* effect at another level.

Above all, the qualitative approach to understanding nature requires more thought about nature itself than about the model details; it is relatively uninterested in precision; and it is *always* represented by more than one model.

Limitations of Models

Models are constructed in order to understand the world. But they are human constructs, not photographs of "reality." They pick out those aspects of the world that we already have decided are important. We manipulate the model to answer the questions we have selected. In constructing a model we choose to ignore some aspects of reality; we group phenomena, we simplify mathematical relations, and we treat things that are inseparable as being distinct aspects of nature. We choose to handle some variables as if they are constant and some changing variables as if they are at equilibrium; we handle complex but determined processes as if they are random, and random processes as if they follow their average behavior. All these typical procedures misrepresent nature in certain ways in order to understand other aspects better.

Therefore, all models are misleading in the long run, although some models are helpful along the way. To rage at things left out or distorted in our models is not helpful. Rather, we must remember that models are our constructions —not the natural objects.

Modelers always must keep in mind that the utility of their construct depends on the particular purpose for which it was built. There is no such thing as *the true model* of a system but only more or less adequate representations of aspects of the system.

The development of models of population growth is instructive. The early study of population growth used a logarithmic model in which population size at a given time, $\tau + G$, is some multiple of the population at a previous time, τ:

$$N(\tau + G) = RN(\tau)$$

where G is generation time. A continuous model gives *instantaneous* rates of change over time t:

$$\frac{dN}{dt} = rN$$

where N is the population size, and R and r the parameters for the intrinsic rate of growth of the discrete and continuous models, respectively.

This model was applied to the growth of bacteria in laboratory culture. If the solution

$$N = N_0 e^{rt}$$

is plotted on a semilog scale the result is a straight line with slope r. If r is positive the population increases, and if r is negative the population decreases. But the real growth of bacteria is not logarithmic. There is an initial lag period before logarithmic growth, and if a culture is kept long enough its growth rate slows down. Nonetheless, the first simple model continues to be applied, although it is restricted to the logarithmic phase. That is, the object of study may be chosen to fit the model.

Real population growth does not continue indefinitely. If our objective is to build this additional reality into the model several choices are available. One common approach is to say that the population growth, r, decreases as some threshold is approached. The familiar logistic equation is

$$\frac{dN}{dt} = r_0 N\left(1 - \frac{N}{K}\right)$$

The previous constant r has been replaced by $r_0(1 - N/K)$ and now there are two parameters, the initial or intrinsic rate of increase, r_0, and the carrying capacity, K. (Chapter 6 discusses differential equations in more detail.)

The new model corrects the failure of the previous equation by allowing the growth of the population to decrease and the population to reach an asymptote. To that extent it represents a conceptual advance. Furthermore, the introduction of the two constants r and K allows us to distinguish readily between initial growth and final level. Rather than seeing populations either as "successful" (r with a large value) or "unsuccessful" (r with a small value), we now can recognize that populations may expand rapidly but cease doing so at a low level, or grow slowly but reach higher limits.

The model, however, is misleading. With a positive value of r (that is, favorable environment prior to density effects) the population expands when below K and decreases when above. But if r is negative (unfavorable physical conditions), and in addition there is overpopulation ($N > K$), the equation tells us that the population will increase! This nonsense result is the consequence of taking the equation too seriously, of applying it to circumstances it was not designed to represent. The model is misleading in another way. The fact that we use the symbols r and K to represent initial slope and final level does not make r and K distinct biological properties determined by different genes and selected for separately. To imagine that our symbols are real-world objects is to reify mathematical constructs into natural objects, a frequent error in model building.

The growth model described above is insufficient in other ways as well: it does not allow for population fluctuations, it treats all individuals as equivalent without age distribution that affects r, and it ignores the way in which population growth becomes self-limiting by way of other variables —resource depletion, or an increase of predators, for example. Therefore the models of population have evolved in a number of different directions.

The important point is that at each stage (logarithmic, logistic, et cetera) the model serves to make some phenomena more understandable while creating new sources of error. The art of model building consists in knowing when a simplification continues to promote understanding and when it outlives its usefulness to become an oversimplification that obscures more than it reveals.

We would like "to work with manageable models which *maximize generality, realism, and precision* toward the overlapping but not identical goals of *understanding, predicting, and modifying nature*" (Levins, 1966, p. 422; emphasis not in original). Unfortunately, it seems to be a fundamental principle of modeling that *no model can be general, precise, and realistic.* Massive predictions bury us in numbers.

Models that are realistic for a particular application may give results in a limited domain. General models may be inaccurate for specific application. From this fundamental principle come three model building strategies:

1. Sacrifice generality for realism and precision. This is a popular approach of resource management (see Holling, 1978; Watt, 1968) and fisheries biologists (Cushing, 1975). These workers can reduce the parameters to those relevant to the short-term behavior of their organisms, make fairly accurate measurements, solve their problems numerically on the computer, and end with precise, testable predictions applicable to the resource management problem.

2. Sacrifice realism for generality and precision. This approach is popular among physicists who enter ecology. Representative of this work are the models of Kerner (1957) and Leigh (1975). Very general equations patterned after statistical mechanics give precise results. But the equations are unrealistic, and even small departures from the assumed conditions have large effects on the outcome.

3. Sacrifice precision for generality and realism. This approach is favored by MacArthur (1972) and Levins (1974) and is the one used in this book. In the long run we really are concerned with qualitative rather than quantitative results. The quantitative results are important in testing hypotheses and in resource management. The models are quite flexible; they only assume functions that are increasing or decreasing, convex or concave, greater or less than some value, and usually do not specify the exact mathematical form of the equations.

As proponents of the third method of model building strategy, we hasten to point out that even the most flexible models are based on artificial assumptions. There is always room for doubt as to whether the conclusions of a model depend on the essentials of the model or on the details of the simplifying assumptions. We restate, therefore, the need for alternative models in order to reach robust conclusions.

Strategy for Qualitative Model Applications

The mathematical approaches taken in this book are not especially advanced but are likely to be unfamiliar because they are not in main-line mathematics education. They all depend in one way or another on the interacting variables. Loop models are special types of graphs known as signed digraphs. An example is depicted in Figure 1.2. The graph corresponds to a matrix of interactions, and to a system of simultaneous equations (differential or difference). Once these equivalences are established, it is possible to work directly with the rules for manipulation of the graphs.

The equivalence of the graph and the algebraic structure (a matrix) has been developed independently by a number of researchers in different

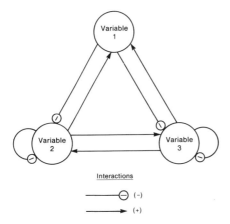

Fig. 1.2

disciplines for very different purposes. Sewall Wright (1921) used the method of path coefficients to calculate the statistical correlations among relatives in different systems of breeding. Samuel Mason (1953) developed it as an algorithm for the computation of gain in electrical circuits. Richard Levins (1974) introduced the procedures of loop analysis for the qualitative interpretation of systems with an intermediate number of variables.

Loop analysis, while related to the other signed digraph methods, places its emphasis on qualitative rather than quantitative predictions—knowing whether variables increase, decrease, or remain unchanged—rather than specifying by precisely how much. The search for a single model is replaced with the search for many models. We can start with the biological knowledge to build a signed digraph, or obtain the data and see which digraphs are consistent with them. We will find many cases when we do not know all the connections between variables; it will be necessary to speculate and use our imagination in building a number of models. Our consolation will be that nature may be more complicated than we like but not as complicated as it could be.

2 Loop Models: Definition of Terms and Calculation of Model Properties

 In this chapter signed digraphs—loop models—are introduced without explanation. We concentrate on the definition of terms and method of analysis.

Symbols of Signed Digraphs

Consider the very simple relation between a predator and its prey, as illustrated in Figure 2.1.

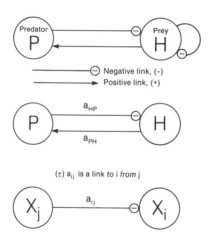

Fig. 2.1

The symbol → (arrow) signifies a positive or enhancing effect on the number of predators, and —⊖ (small circle enclosing a dash) represents a negative or decreasing effect on the number of prey. The picture is readily interpreted to mean that the predator consumes the prey, thus negatively affecting the prey; while the prey, by being consumed by the predator, have a positive effect on the predator. The lines with small arrowheads or small circleheads are called *links*. The links connect the *variables*, which are shown by the big circles and form the vertices of a graph. The variables are the objects that make up the system we are studying. One link of special importance is that which connects a variable to itself, the self-effect link. With a small circlehead, as shown, this is the *self-damping* of a variable. If there had been an arrowhead on the link, then it would be called *self-enhancing*, or autocatalytic.

Self-effects come about either because the growth rate of a variable depends on its own density (amount per unit area) or because there is a source for a continuous supply of the variable from outside the system being modeled; hence, a self-effect can be independent of its own density. A self-effect is generally positive if a constant number of individuals is removed per unit time, as in hunting quotas. Self-effects are absent when the growth of a species is proportional to its own abundance, so that the *rate* per individual is independent of its own abundance; there is no outside supply and the growth rate can depend only on other variables. Of course, variables like species do affect their own growth rates by using up resources, by contaminating their environment, or by affecting their predators. But these effects are by way of other variables in the system. If these other variables are included in the model, the self-link would be omitted; but if a variable acts on its own growth rate by a pathway that is not represented explicitly by other variables, then the self-loop would be necessary. To some extent it is a matter of choice whether to recognize an intermediate variable along the pathway from one variable to itself. Usually, if the intermediate variable has no other connections and is produced and removed rapidly, in comparison to the demographic events, there is no need to represent it directly. Thus, for example, aggregate interactions require at least a temporary formation of interacting pairs, but we usually incorporate this into a self-inhibition loop.

The notion of self-effects may be easier to understand on the basis of mathematical formulation. For now it is simpler to take as given that most nutrients in an ecosystem are self-damped; or, when these are omitted in the model, the primary producers and lowest levels of the trophic structure become self-damped. In general, although not always, the top level of the trophic structure is not self-damped. Exceptions to this general rule typically occur because nonfood items—space, nutrients, the input of a younger age-class from outside the system boundaries, cannibalism—also influence the growth rate of species. In biochemical networks the variables are not self-reproducing; each chemical arises from some precursor. Therefore, the chemicals are self-damped. Similarly, the firing of a neuron is a self-damped variable.

Notation

Each link can have a symbol assigned to it, to represent the unknown magnitude of the interaction. In the predator–prey example we can denote the links as indicated in Figure 2.1. The notation is a letter from the alphabet (roman or greek) with a pair of subscripts. The first subscript denotes the end point of the link and the second denotes the point of origin. Therefore, $\pm a_{ij}$ signifies a link to i from j. The sign associated with the symbol depends on whether the link has a single arrowhead or small circlehead associated with it, and indicates the effect on i from j.

It is generally obvious from a description of a system what the links between variables should be. In the last section of Chapter 3 and in the discussion on self-effect terms in Chapter 6, however, we expand the intuitive notion of interconnections between variables and give a formal mathematical rule for calculating the links. We strongly emphasize that exercising *only* common sense about interconnections can be wrong—sometimes links between variables are included erroneously or, as will happen most often, necessary links between variables are excluded.

Paths, Links, and Loops

A *path* is defined as a series of links starting at one variable and ending on another, without crossing any variable twice. For example, Figure 2.2 is a system of three variables: a nutrient, a herbivore, and a predator.

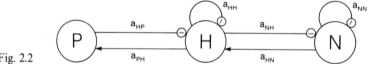

Fig. 2.2

The path P to H is

$$-a_{HP}$$

but not

$$a_{HP} a_{NH} a_{HN}$$

The path from P to N is

$$a_{HP} a_{NH}$$

Strictly speaking, it is $(-a_{HP})(-a_{NH})$, but the multiplication becomes positive, so we write $+a_{HP}a_{NH}$ and drop the plus sign.

Paths can have one or many links. The number of links determines the path length. A path length is one less than the number of variables along the path. There can be more than one path from one variable to another, as illustrated in Figure 2.3.

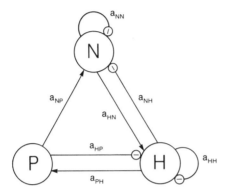

Fig. 2.3

In this example the predator P secretes a nutrient N consumed by the herbivore H. The path from P to N is either the long pathway

$$a_{HP}a_{NH}$$

or the direct single link path

$$a_{NP}$$

A path that returns to its starting point, not crossing any intermediate variable twice, is called a *loop*. In Figure 2.3, going from P to N to H to P is a loop. In link notation, this loop can be written as

$$(a_{NP}a_{HN}a_{PH}) \quad \text{or} \quad [PNH]$$

This is a loop of length 3 because it has three links and three variables. Notice that a loop length has the same number of links as variables, unlike the path length, in which the number of links is one less than the number of variables. A loop begins and terminates on the same variable, so this variable does not add twice to the variable numbers; yet there are two links associated with it.

Note: A self-effect link is defined as a loop of length 1.

Most systems have more than one loop, of either the same or different lengths. In the simple predator-prey example stated at the beginning of the chapter and repeated in Figure 2.4, there are just two loops: the loop of length 1, $(-a_{HH})$. and the loop of length 2 $(-a_{HP}a_{PH})$.

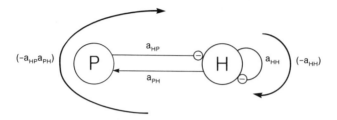

Fig. 2.4

The three-variable model, repeated in Figure 2.5, has the following loops:

Loops of length 1:
 (i) $(-a_{NN})$
 (ii) $(-a_{HH})$

Loops of length 2:
 (i) $(-a_{HN}a_{NH})$
 (ii) $(-a_{HP}a_{PH})$

Loops of length 3:
 (i) $(a_{HN}a_{PH}a_{NP})$

Notice that each subscript in a loop appears twice, once as a first subscript indicating input to a variable, and once as a second subscript indicating output.

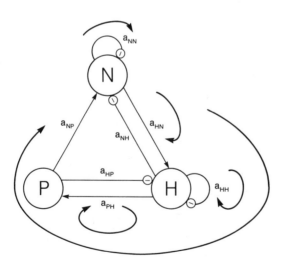

Fig. 2.5

Conjunct and Disjunct Loops

The loops in the same system are either *conjunct* or *disjunct*. *Conjunct loops* are loops that have at least one variable in common. *Disjunct loops* have no variables in common. In the example of Figure 2.5 the loops of length 2 are conjunct because they each include the herbivore H. The predator, prey, and nutrient model in Figure 2.5 is an example of a system that has both conjunct and disjunct loops.

Conjunct
$(-a_{NN})$ and $(-a_{HN}a_{NH})$
$(-a_{NN})$ and $(a_{NP}a_{HN}a_{PH})$
$(-a_{HH})$ and $(-a_{HN}a_{NH})$
$(-a_{HH})$ and $(-a_{HP}a_{PH})$
$(-a_{HH})$ and $(a_{NP}a_{HN}a_{PH})$
$(-a_{HN}a_{NH})$ and $(-a_{HP}a_{PH})$
$(-a_{HN}a_{NH})$ and $(-a_{NP}a_{HN}a_{PH})$
$(-a_{HP}a_{PH})$ and $(a_{NP}a_{HN}a_{PH})$

Disjunct
$(-a_{NN})$ and $(-a_{HH})$
$(-a_{NN})$ and $(-a_{HP}a_{PH})$

Feedback

Feedback is defined in terms of *disjunct loops*. Bear in mind that loops may differ in length, and disjunct loops have no variables in common.

As commonly used, the term "feedback" denotes that an action or activity initiated by someone or something sets in motion activities or responses by others which then affect the original source of the activity. We are going to turn this general idea of feedback into a mathematical operation on loop diagrams.

Before proceeding to the mathematics of feedback calculations we should consider in some tangible way what feedback does. To do so we must recognize two types of feedback, positive and negative. *Positive feedback* occurs when an *increase* in one variable (the initiator) causes other variables to change in such a way as to *increase itself* (the initiator) *further*; or, conversely, if a variable (the initiator) *declines*, this creates conditions for further *decrease*. Any variable along the loop may be the initiator, with the same result. Two contrasting examples of positive feedback follow. (1) A politician makes friends in the political caucus, which causes her to get elected. Holding political office enables the politician to make more friends in the political arena, which increases the likelihood of reelection. (2) If teachers treat children as poor learners, learning is inhibited, and the negative attitude of the teacher is reinforced.

Negative feedback occurs when an *increase* in a variable (the initiator) sets in motion processes that lead to a *decrease* in that variable (the initiator); or, conversely, a *decrease* in a variable leads to processes that *increase* the variable (the initiator). Here are two contrasting examples. (1) A baby cries when it is uncomfortable, causing adults to respond by removing the cause of discomfort; the crying stops. (2) A decline in trout at a lake resort triggers a series of events which cause a governmental department of fish and wildlife protection to take actions that increase fish abundance.

Two further points about feedback should be noted. First, the qualifiers "positive" and "negative" are not value judgments. In the physical and biological sciences, negative feedback contributes to a system being stable. Insofar as many people think stable systems are desirable, however, negative feedback would seem a good thing. For social scientists, positive feedback means that someone receives a pleasant response from others for whatever that individual has done; negative feedback means that someone's efforts have not been appreciated. Social scientists would view negative feedback as the consequence of something undesirable. We reiterate: positive and negative feedback in loop models are not value judgments.

The second point about feedback is that in the definitions and examples stated above, any number of things or people can be involved in a process or action, which then affects the initiating variable. Moreover, many pathways can be taken for the same starting event. For example, when the fish population in the lake declines this could set into action complaints by the local chamber of commerce to the state regulatory agency, which then stocks more fish; it might simultaneously cause the local fishermen's association to pressure the town officials to take action. They, in turn, may increase fishing license fees for outsiders, thereby reducing the number of people who fish.

Any system will have multiple feedback paths, not necessarily of equal length. Consequently, we will introduce the notion of *feedback level*, the number of variables that are part of the feedback path. The feedback levels in any system will range from one to the total number of variables in the system.

We now are ready to introduce the calculation of feedback for loop diagrams. Feedback at any level, say k, will be denoted as F_k. Feedback at level 0 is defined for all systems as

$$F_0 \equiv -1$$

(This convention will be explained in Chapter 6.)

Let us examine feedback for the hypothetical loop diagram of Figure 2.6. A general formula for any system will be given at the end of the discussion.

Table 2.1 summarizes the feedback at each level for Figure 2.6.

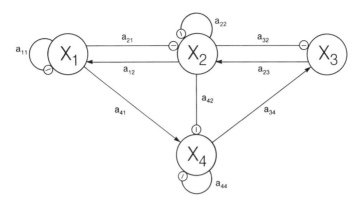

Fig. 2.6

The feedback calculation of the model in Figure 2.6 highlights the following. First, feedback at higher levels is composed of combinations of feedback at lower levels plus longer pathways. Feedback at Level 4 includes the long pathway of four variables—$X_1 X_4 X_3 X_2$—and also combinations of products of pathways which occurred at Level 3, such as $[X_1][X_2 X_4 X_3]$. Likewise feedback at Level 3 includes length 3 pathways and combinations of pathways found at feedback Level 2. Second, only disjunct loops appear in the products. We do not include, for instance, $(-a_{11})(-a_{12}a_{21})$, that is, $[X_1][X_1 X_2]$, in the Level 3 feedback, because this is a product of conjunct loops.

In general, feedback for any loop model can be computed as follows. Feedback at level k is found by determining all the loops of length k or product of disjunct loops that have a combined length of k, and then adding them together. In formula this is

$$F_k = \sum_{m=1}^{k} (-1)^{m+1} L(m, k)$$

where $L(m, k)$ is the notation meaning m disjunct loops with k elements. For example, $L(1, 4)$ signifies one disjunct loop of length 4; there are four variables in a single loop. $L(2, 4)$ signifies two disjunct loops whose total length is 4. This could mean one loop of length 3 and the other of length 1, or both loops of length 2.

The $(-1)^{m+1}$ tells us that if the number of disjunct loops multiplied together is even, then their product is multiplied by -1. Conversely, an odd number of disjunct loops does not alter the product sign. This rule is useful because we can see at a glance that any set of variables of all negative loops contributes a negative feedback term.

The number of disjunct loops cannot exceed the sum of the length of all the

Table 2.1 Feedback calculation for Figure 2.6

Level ($k=$)	Verbal Description	Feedback	Explanation	Variables
0	Definition	$F_0 = -1$	See Chapter 6	
1	*Addition* of all loops of length 1	$F_1 = (-a_{11})$ $+ (-a_{22})$ $+ (-a_{44})$	A loop of length 1 + A loop of length 1 + A loop of length 1	$[X_1]$ $[X_2]$ $[X_4]$
2	*Addition* of all loops of length 2 AND *subtraction* of all combinations of pairwise products of disjunct loops of length 1	$F_2 = (-a_{12}a_{21})$ $+ (-a_{23}a_{32})$ $- (-a_{11})(-a_{44})$	A loop of length 2 + A loop of length 2 − A product of loops of length 1	$[X_1X_2]$ $[X_2X_3]$ $[X_1][X_2]$
3	*Addition* of all loops of length 3 AND *subtraction* of all combinations of pairwise products of disjunct loops of length 1 with loops of length 2 AND *addition* of all combinations of triplet products of disjunct loops of length 3	$F_3 = (-a_{42}a_{34}a_{23})$ $- (-a_{11})(-a_{23}a_{32})$ $- (-a_{44})(-a_{12}a_{21})$ $- (-a_{44})(-a_{23}a_{32})$ $+ (-a_{11})(-a_{22})(-a_{44})$	A loop of length 3 − A product of loop of length 1 with a loop of length 2 + A product of loop of length 1 with a loop of length 2 − A product of loop of length 1 with a loop of length 2 + A product of loops of length 1	$[X_2X_4X_3]$ $[X_1][X_2X_3]$ $[X_4][X_1X_2]$ $[X_4][X_2X_3]$ $[X_1][X_2][X_4]$

4	Addition of all loops of length 4 AND subtraction of all combinations of pairwise products of disjunct loops of length 1 with loops of length 3 AND subtraction of all combinations of pairwise products of disjunct loops of length 2 AND addition of all combinations of triplet products of two disjunct loops of length 1 with one disjunct loop of length 2 AND subtraction of all combinations of quadruplet products of disjunct loops of length 1	$F_4 = (+a_{41}a_{34}a_{23}a_{12})$ $-(-a_{11})(-a_{42}a_{34}a_{23})$ -0 -0	A loop of length 4 — A product of a loop of length 1 with a loop of length 3 — Combination of products of loops of length 2 with loops of length 2 + Combination of products of four loops of length 4	$[X_1 X_4 X_3 X_2]$ $[X_1][X_2 X_4 X_3]$ NONE NONE
5	Cannot be calculated for this system	$F_5 = ?$	This hypothetical example does not have five variables	Not enough variables
6	Cannot be calculated for this system	$F_6 = ?$	This hypothetical example does not have six variables	Not enough variables
⋯	⋯	⋯	⋯	⋯

loops, such as $L(5, 4)$; you cannot get five disjunct loops whose total length is only four—that is, only four variables. Even if there are five disjunct loops, each with only the minimum length 1, it would give a total length of 5.

EXAMPLE 2.1

Calculate feedback for all levels of a simple predator-prey system, as given in Figure 2.7.

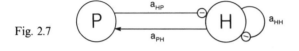

Fig. 2.7

Solution

$$F_1 = (-a_{HH})$$
$$F_2 = (-a_{HP}a_{PH})$$

EXAMPLE 2.2

Calculate feedbacks for a three-level system of predator, herbivore, and nutrient, as depicted in Figure 2.8.

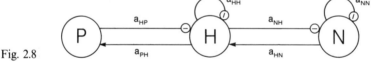

Fig. 2.8

Solution

$$F_1 = (-a_{HH}) + (-a_{NN})$$

$$F_2 = \sum_{m=1}^{2} (-1)^{m+1} L(m, 2)$$

$$= (-a_{HP}a_{PH}) + (-a_{NH}a_{HN}) \quad \boxed{m = 1 \text{ (one disjunct loop)}}$$

$$+ (-1)(-a_{HH})(-a_{NN}) \quad \boxed{m = 2 \text{ (two disjunct loops)}}$$

$$F_3 = \sum_{m=1}^{3} (-1)^{m+1} L(m, 3)$$

= NONE $\quad$ | $m = 1$ (one disjunct loop) |

$+ (-1)(-a_{HP}a_{PH})(-a_{HH})$ $\quad$ | $m = 2$ (two disjunct loops) |

+ NONE $\quad$ | $m = 3$ (three disjunct loops) |

□

EXAMPLE 2.3

(a) Change the model of Figure 2.8 so that there are three disjunct loops at Level 3.

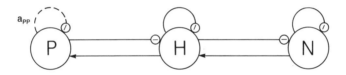

(b) Change the model so there is one disjunct loop at Level 3.

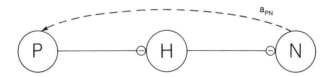

Solution

(a) Add self-damping to the predator.

$$F_3 = (-a_{PP})(-a_{HH})(-a_{NN})$$

(b) Include a link from the nutrient to the predator. □

Stability

FIRST EXPLANATION

The variables of our system—be they species, nutrients, or anything else—have a quantity associated with them: species number, biomass, or concentration. If these quantities are at some level (that is, have a value) and then something happens to the system, we can wonder what happens to these values. Do the levels of the variables change? When levels increase or decrease and then return to their previous levels, we think of this as a stable system. When levels increase or decrease unbounded or oscillate forever, never to return to their original levels, we think of this as an unstable system. This is, of course, a rather simplified, naive explanation of stability in the neighborhood of equilibrium, but it does describe the kinds of question we will discuss concerning stability.

To be specific, how can we determine whether the system we are studying is stable or unstable? Perhaps more to the point is the question, what are the properties of the system that tend to stabilize or destabilize it?

LOOP MODEL CRITERIA FOR LOCAL STABILITY

The stability criteria based on the loop model notation are twofold. First, there is the condition that feedback at all levels be negative. By "all levels" we mean from Level 1 up to the maximum possible, which is the total number of variables in a system. This can be expressed in a mathematical statement as

$$F_i < 0 \quad \text{for all } i$$

The second condition relates the negative feedback of long loops to the negative feedback of short loops. It asserts that negative feedback at high levels cannot be too strong, compared to lower levels. We will defer giving the full algorithm for this criterion until Chapter 6, but note that the first case of interest relates feedback up to Level 3. Then the second criterion for three variables is expressed as

$$F_1 F_2 + F_3 > 0$$

For four variables the same formula applies. For five variables the second condition for stability is

$$-(F_1 F_2 + F_3)F_3 + (F_1 F_4 + F_5)F_1 > 0$$

or

$$F_1^2 F_4 + F_1 F_5 - F_1 F_2 F_3 - F_3^2 > 0$$

For more variables the formula of the second condition gets more complicated.

EXAMPLE 2.4

Calculate the stability conditions for the simple predator–prey model at the beginning of the chapter.

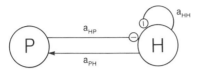

Solution

Stability Condition 1: $F_i < 0$, for all i.

$F_1 < 0$ ✓ $(F_1 = -a_{HH})$

$F_2 < 0$ ✓ $(F_2 = -a_{HP}a_{PH})$

∴ Stable □

Note: Quantification does not add anything to this conclusion.

EXAMPLE 2.5

Calculate the stability criteria for the three-variable, linear chain model.

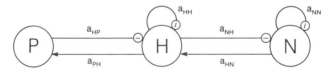

Solution

Stability Condition 1: $F_i < 0$, for all i.

$F_1 < 0$ ✓ $(F_1 = -a_{HH} - a_{NN})$

$F_2 < 0$ ✓ $(F_2 = -a_{HP}a_{PH} - a_{NH}a_{HN} - a_{HH}a_{NN})$

$F_3 < 0$ ✓ $(F_3 = -(a_{NN})(a_{HPPH}))$

Stability Condition 2: $F_1F_2 + F_3 > 0$.

$\{(-a_{HH}) + (-a_{NN})\}\{(-a_{HP}a_{PH}) + (-a_{NH}a_{HN}) + (-a_{HH}a_{NN})\}$
$+ \{-(a_{NN})(a_{HP}a_{PH})\} > 0$ ✓

$= + a_{NH}a_{HP}a_{PH} + a_{HH}a_{NH}a_{HN} + a_{HH}^2 a_{NN}$
$ + \cancel{a_{NN}a_{HP}a_{PH}} + a_{NN}a_{NH}a_{HN} + a_{HH}a_{NN}^2$
$ \cancel{- a_{NN}a_{HP}a_{PH}}$

$\rule{4cm}{0.4pt}$

> 0 □

EXAMPLE 2.6

Calculate the stability criteria for the three-variable triangular system.

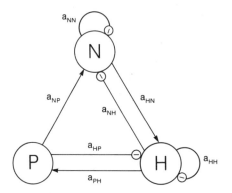

Solution

Stability Condition 1: $F_i < 0$, for $i = 1$ to 3.

$F_1 < 0 \quad [F_1 = -a_{HH} - a_{NN}]$

$F_2 < 0 \quad [F_2 = -a_{HP}a_{PH} - a_{HN}a_{NH} - (a_{HH})(a_{NN})]$

$F_3 < 0 \quad [F_3 = +a_{PH}a_{NP}a_{HN} - (a_{NN})(a_{HP}a_{PH})]$

$F_3 < 0$ if $|a_{NN}a_{HP}a_{PH}| > |a_{PH}a_{NP}a_{HN}|$

(Short loops are stronger than long loops.)

Stability Condition 2: $F_1F_2 + F_3 > 0$.

$\{-(a_{HH} + a_{NN})\}\{-(a_{HP}a_{PH} + a_{HN}a_{NH} + a_{HH}a_{NN})\}$
$\quad + \{a_{PH}a_{NP}a_{HN} - a_{NN}a_{HP}a_{PH}\}$
$= a_{HH}a_{HP}a_{PH} + a_{HH}a_{HN}a_{NH} + a_{HH}^2 a_{NN}$
$\quad \cancel{+ a_{NN}a_{HP}a_{PH}} + a_{NN}a_{HN}a_{NH} + a_{HH}a_{NN}^2$
$\quad \cancel{- a_{NN}a_{HP}a_{PH}} + a_{PH}a_{NP}a_{HN}$
$= + > 0 \qquad \square$

Stability is guaranteed, provided the assumption of the relative strengths of the short loops to long loops hold.

Note that when distinct feedbacks are multiplied, conjunct loops enter the

LOOP MODELS: TERMS AND PROPERTIES

calculation; the term $(a_{HP}a_{PH})a_{HH}$ is a product of two loops with the vertex H in common. As we see in Figure 2.9, the system is unstable because

$$F_1F_2 = (-a_{AA})[a_{BC}(-a_{CB})]$$
$$F_3 = -a_{AA}a_{BC}(-a_{CB}) + a_{BA}a_{CB}(-a_{AC})$$
$$F_1F_2 + F_3 < 0$$

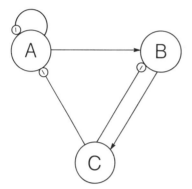

Fig. 2.9

If the loop of length 2 is transferred, as shown in Figure 2.10, we do not know if it is locally stable.

Local stability will depend on whether

$$(-a_{AA})[a_{BA}(-a_{AB})] > a_{BA}a_{CB}(-a_{AC})$$

or

$$a_{AA}a_{AB} > a_{CB}a_{AC}$$

Thus the positions of loops in the graph can influence the outcome. Notice also that a_{BA} does not affect local stability at all.

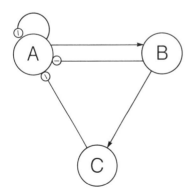

Fig. 2.10

It often is claimed that negative feedback stabilizes a system, whereas positive feedback destabilizes. This is not necessarily so, however. We already have seen from the second stability criterion (the Routh-Hurwitz criterion), that

$$F_1 F_2 + F_3 > 0$$

and that strong negative feedback coming from loops of length 3 can cause oscillatory instability. In the examples that follow we see how the effect of a particular loop depends on the rest of the system, and that a loop may destabilize in one context and stabilize in another.

EXAMPLE 2.7

Consider the system below, where the positive feedback may arise if X is a harvestable species and a fixed number of individuals is removed per unit time. Show that a small amount of positive feedback does not destabilize the system.

Solution
The first criterion for stability is that $F_i < 0$ and therefore

$$-a_{NN} + a_{XX} < 0 \quad \text{or} \quad a_{NN} > a_{XX}$$

At Level 2,

$$F_2 = +a_{NN} a_{XX} - a_{NX} a_{XN} < 0$$

But since a_{XX} is positive, a_{NN} contributes to positive feedback at Level 2. We require

$$a_{NN} < \frac{a_{NX} a_{XN}}{a_{XX}}$$

Combining the two conditions,

$$a_{XX} < a_{NN} < \frac{a_{NX} a_{XN}}{a_{XX}}$$

That is, the negative feedback of N to itself must be stronger than the self-excitation of X, but not too much stronger. And if

$$a_{XX}^2 > a_{NX}a_{XN}$$

the conditions for stability can never be satisfied, no matter what the self-damping of N may be. □

It also is possible for positive feedback to have a stabilizing effect within certain boundaries.

EXAMPLE 2.8

Consider the system below and show that a certain amount of positive feedback is necessary for stability.

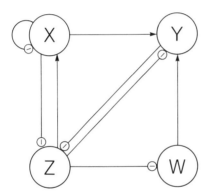

Solution

Feedbacks at Levels 1 and 2 are negative, provided predator–prey interactions are stronger than competitive ones. At Level 3, stability requires that

$$-(-a_{XX})(a_{YZ}a_{ZY}) + (-a_{YX}a_{ZY}a_{XZ}) + (a_{YW}a_{ZY}a_{WZ}) < 0$$

Therefore the positive feedback cannot be too strong. But the second criterion for stability (Routh–Hurwitz) requires that

$$(-a_{XX})(-a_{XZ}a_{ZX}) + (-a_{YX}a_{ZY}a_{XZ}) + (a_{YW}a_{ZY}a_{WZ}) > 0$$

Therefore stability requires

$$(a_{YW}a_{ZY}a_{WZ}) + (a_{XX})(a_{XZ}a_{ZX}) > (a_{YX}a_{ZY}a_{XZ}) > (a_{YW}a_{ZY}a_{WZ}) \\ + (a_{XX})(a_{YZ}a_{ZY})$$

and the long positive feedback loop ($a_{YW}a_{ZY}a_{WZ}$) cannot be too weak. Note that this relation can only be satisfied when

$$a_{YZ}a_{ZY} < a_{XZ}a_{ZX}.$$

Finally, because of the way loops are multiplied in the second stability calculation, the product of two negative disjunct loops contributes to positive feedback. □

EXAMPLE 2.9

Show that in the system given below a certain amount of positive feedback is necessary for stability.

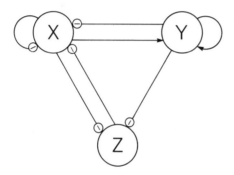

Solution

At Level 1 we require $-(-a_{XX}) > (a_{YY})$. At Level 2,

$$(a_{XZ}a_{ZX}) + (-a_{XY}a_{YX}) - (-a_{XX})(a_{YY}) < 0$$

so that

$$a_{XZ}a_{ZX} < -a_{XX}a_{YY} + a_{XY}a_{YX}.$$

At Level 3,

$$-(a_{YY})(a_{XZ}a_{ZX}) + (a_{YX}a_{ZY}a_{XZ}) < 0$$

or

$$a_{XZ}a_{ZX} > \frac{a_{YX}a_{ZY}a_{XZ}}{a_{YY}}$$

Therefore,

$$-a_{XX}a_{YY} + a_{XY}a_{YX} > a_{XZ}a_{ZX} > \frac{a_{YX}a_{ZY}a_{XZ}}{a_{YY}} \tag{E.2.9.1}$$

Thus the positive feedback $(a_{XZ}a_{ZX})$ cannot be too weak or the system will be unstable. The inequality of Equation E2.9.1 also requires that

$$-a_{YY}^2 a_{XX} + a_{YY}a_{XY}a_{YX} > a_{YX}a_{ZY}a_{XZ}$$

or the conditions on $(a_{XZ}a_{ZX})$ cannot be satisfied.

The final constraint, $F_1 F_2 + F_3 > 0$, places another upper bound on $(a_{XZ}a_{ZX})$. Since

$$[-a_{XX} + a_{YY}][-a_{XY}a_{YX} + a_{XZ}a_{ZX} + a_{XX}a_{YY}]$$
$$+ [a_{YX}a_{ZY}a_{XZ} - a_{YY}a_{XZ}a_{ZX}] > 0$$

we require that

$$(a_{XZ}a_{ZX}) < [a_{XY}a_{YX} - a_{XX}a_{YY}]$$
$$+ \frac{-a_{YY}a_{XZ}a_{ZX} + a_{YX}a_{ZY}a_{XZ} + a_{YY}^2 a_{XX}}{a_{XX}}$$

The first term on the right is the upper bound on $(a_{XZ}a_{ZX})$ from Equation E2.9.1 The second term may be positive or negative, so that the last equation could make the stability requirement on $(a_{XZ}a_{ZX})$ more stringent if $-a_{YY}a_{XY}a_{YX} + a_{YX}a_{ZY}a_{XZ} + a_{YY}^2 a_{XX} < 0$. □

3 Loop Models: Predicting Change

We have shown how signed digraphs—loop models—enable us to represent variables (for example, animal populations) as part of a network of interacting variables. Given a loop model, we sometimes can calculate whether the system is locally stable or not. If we cannot determine the stability, we can identify the measurements necessary to decide.

In this chapter we develop qualitative analysis so we can predict how variables respond to changed conditions. The following section provides some basic definitions and focuses our attention on what can be learned from change. The mathematical details are in the last section of the Appendix.

Parameter Change

Understanding the mechanism of parameter change underscores what the analysis of loop models is all about. Human alteration or impingement on an ecosystem demonstrates this very well, as in the case of spraying crops to control weeds. The herbicides increase the mortality parameter of the weeds and, perhaps, other parameters. There is a direct change in the parameters of the equation for rate of change in the weed population.

Natural changes, like human action, are changes in the parameters of the system. A warming trend increases the growth rate of the weeds and the crop plants. The weather produces changes in parameters of two of the variables of the crop ecosystem.

What would happen to the amount of weeds or crops as a result of either the herbicide spray or general warming trend is the subject of predicting change. We will develop this qualitative analysis using the simple crop–weed–herbivore system of Figure 3.1.

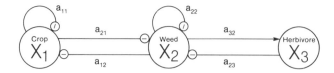

Fig. 3.1

As we see in this figure, the weed and crop are in competition for space and nutrients. Both the crop and weed depend upon the nutrients and water in the soil, which become available at a rate independent of their levels; hence, they are self-damped. Since the model does not deal directly with the nutrients or water, their self-damping property is transferred to the plants. Growth of the herbivore population depends on the amount of weeds eaten per individual and has a natural rate of mortality; thus, there is no self-damping.

Consider the simple case of spraying a herbicide. This step only changes the rate of mortality of the weeds. This is called a *parameter change*. Such changes take place when there is an event occurring outside the system which causes a change in one or more of the parameters of a variable. In this case, an increase in the mortality rate parameters are represented in the loop model as a link attached only to the variable whose parameters are changed. We use the convention of a small circlehead to signify that the change decreases the growth rate of the variable. An arrowhead signifies the growth rate increases by the change in the parameters (Fig. 3.2).

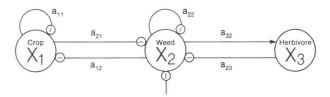

Fig. 3.2

Figure 3.2 stresses that the direct change appears only in the weed equation. Nevertheless, because of the interconnection of weeds to the crop and herbivores there may be effects beyond the local entry point of the parameter change. We will demonstrate how to calculate these effects and understand the properties of a network of interactions.

In this chapter we are analyzing changes brought about by changing the parameters of a given system. It is difficult to think of cases, at least in the human intervention category, where things can change yet do not involve changes of the parameters. The reason is simple: human activity, such as thermal discharge from electric power generation, oil spills, dredging, interferes with the interaction of the species in the system. The interaction is represented by the parameters of the system. Hence, oil in the marine environment which reduces fish chemoreceptivity reduces its feeding rate of zooplankton.

In addition to parameter change there is structural change. In this type of change, the interconnections between variables are altered and links are added and/or removed in the corresponding loop model. Or, variables may be added or removed from the entire system.

Any time there is a structural change we cannot predict the new links from the qualitative model; rather, we must determine them by independent methods. An example of a structural change is when the mandate for a state regulatory agency to control the lake fish population via employing stock replacement is altered to a mandate to control the population through levying fines and fixing catch limits on the fishing industry.

From the modeling point of view, we can give substance to the notion of parameter change by making a closer examination of the mathematics involved. (Those unfamiliar with the mathematics are referred to Chapter 6 and should skip this example.)

EXAMPLE 3.1

Given a system with a nutrient N and a consumer H, illustrate how changes in the parameters for consumption rate, p, and assimilation rate, r, alter the values of the equilibrium abundances of N and H.

Solution

Let the relation between the nutrient N and the consumer H be represented at equilibrium by

$$\frac{dN}{dt} = -pHN + I = 0 \tag{E3.1.1}$$

$$\frac{dH}{dt} = H(rN - \theta) = 0 \tag{E3.1.2}$$

So

$$H^* = \frac{I}{pN^*} = \frac{Ir}{p\theta} \tag{E3.1.3}$$

$$N^* = \frac{\theta}{r} \tag{E3.1.4}$$

Let $I = I_0$ and $\theta = \theta_0$ be constants. Then for $p = p_1$, $r = r_1$, we have

$$H^* = \frac{I_0 r_1}{p_1 \theta_0} = H_1^*$$

$$N^* = \frac{\theta_0}{r_1} = N_1^*$$

For a different value of p, $p = p_2$ (assume $p_2 > p_1$, or $|p_2| < |p_1|$ since in

Equation E3.1.1 the p value is negative), we have

$$H^* = \frac{I_0 r_1}{p_2 \theta_0} = H_2^* > H_1^*$$

$$N^* = \frac{\theta_0}{r_1} = N_1^*$$

Thus the consumer population H increases while there is no change in the nutrient level from N_1^*.

For the original p_1 value and a new value of r, $r = r_2$, we get (where we assume $r_2 > r_1$)

$$H^* = \frac{I_0 r_2}{p_1 \theta_0} = H_3^* > H_1^*$$

$$N^* = \frac{\theta_0}{r_2} = N_2^* < N_1^*$$

This condition of changing parameters is plotted in Figure 3.3.

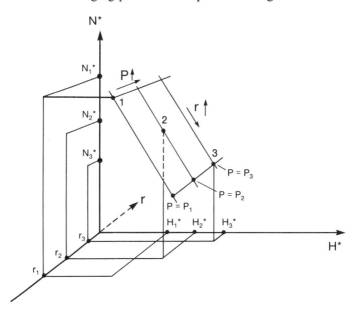

Fig. 3.3

As the parameters change the equilibrium point moves. In the case of parameter r, the larger r becomes, H^* increases accordingly, yet N^* becomes smaller. When parameter p increases in value, so will H^*, with no change in N^*. This is precisely the situation predicted by the loop diagram for Equation 3.2, shown in Figure 3.4.

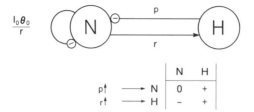

Fig. 3.4

Open Paths and Complementary Subsystems

Before we define two new properties of loop models, ancillary for the algebraic algorithms to be developed, we want to reiterate our purpose. The problem is to predict and understand the effects of parameter changes on the equilibrium values of all the variables of the system. Stated another way, if we ignore the transient response, do the levels of the variables—be they numbers of individuals, amount of biomass, nutrient concentrations, or whatever is being measured—increase, decrease, or remain the same after the parameter change occurs? This is a *qualitative* question. We also can quantify by asking, How much do they change? Quantification requires that the values of the links are known.

To make qualitative predictions using loop models we need to explain additional aspects of signed digraphs.

We previously defined an open path as a series of links starting at one variable and ending on another, without crossing any variable twice. In addition, path *length* is equal to the number of links; this is the same as the number of variables along the path, minus one. The symbol for paths from variable i to j of length $k-1$ is $p_{ji}^{(k)}$. This is depicted in Figure 3.5.

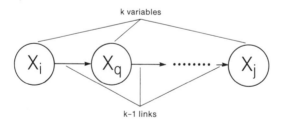

Fig. 3.5

We can have more than one path of the same length to variable i from variable j, as is illustrated in Figure 3.6.

For completeness, we define the value of a path from a variable to itself as *unity*

$$p_{ii}^{(k)} \equiv p_{ii}^{(1)} \equiv p_{ii}^{(0)} \equiv 1$$

k with any value other than 1 makes no sense.

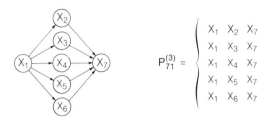

Fig. 3.6

$$P^{(3)}_{71} = \begin{cases} X_1 \ X_2 \ X_7 \\ X_1 \ X_3 \ X_7 \\ X_1 \ X_4 \ X_7 \\ X_1 \ X_5 \ X_7 \\ X_1 \ X_6 \ X_7 \end{cases}$$

EXAMPLE 3.2

Find all the open paths from the nutrient (N) to the carnivore (C), $p^{(k)}_{CN}$, for all lengths in the marine benthic community model of Figure 3.7.

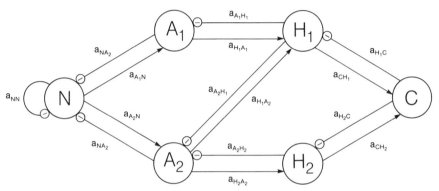

Fig. 3.7 N = nutrient, A_1 = macrophytic algae, H_1 = herbivore, C = carnivore, A_2 = sea grass, H_2 = herbivore.

Solution

Length (k)	Open Path $p^{(k)}_{CN}$	Value (links)
1	none	0
2	none	0
3	none	0
4	$[NA_1H_1C]$	$a_{A_1N}a_{H_1A_1}a_{CH_1}$
	$[NA_2H_2C]$	$a_{A_2N}a_{H_2A_2}a_{CH_2}$
	$[NA_2H_1C]$	$a_{A_2N}a_{H_1A_2}a_{CH_1}$
5	$[NA_1H_1H_2C]$	$a_{A_1N}a_{H_1A_1}a_{H_2C}a_{CH_2}$
	$[NA_2H_1H_2C]$	$a_{A_2N}a_{H_1A_2}a_{H_2H_1}a_{CH_2}$
6	$[NA_1H_1A_2H_2C]$	$a_{A_1N}a_{H_1A_1}a_{A_2H_1}a_{H_2A_2}a_{CH_2}$ □

Given an open path $p^{(k)}_{ji}$ there is a subsystem related to this path, defined as the *complementary subsystem*: the subsystem of variables and their interconnections which are *not* included in the open path. The complementary subsystem then must have $n - k$ variables: The total system has n variables and the number of variables in the path is k, so $n - k$ is the remaining number of variables. The feedback of the complementary subsystem is identical to the formula for feedback for the entire system. It is the feedback at level $n - k$ for the entire subsystem. The symbol used to remind us that it is a complementary subsystem is $F_{n-k}^{(comp)}$. The null subsystem is one which has no variables. This is because all the variables are on the path. We are interested in the paths from the variable whose parameter changes to the variable whose direction of change we want to find. The variable whose parameter changes is the point of

I: Model

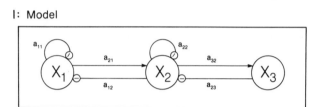

1. Path: $P_{21}^{(1)}$

   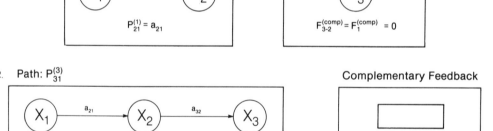

   $P_{21}^{(1)} = a_{21}$

   Complementary Feedback

   $F_{3-2}^{(comp)} = F_1^{(comp)} = 0$

2. Path: $P_{31}^{(3)}$

   $P_{31}^{(3)} = a_{21}a_{32}$

   Complementary Feedback

   $F_{3-3}^{(comp)} = F_0^{(comp)} \equiv -1$

3. Path: $P_{22}^{(1)}$

   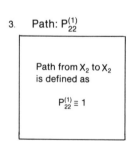

   Path from X_2 to X_2 is defined as

   $P_{22}^{(1)} \equiv 1$

   Complementary Feedback

   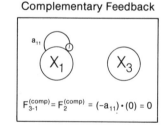

   $F_{3-1}^{(comp)} = F_2^{(comp)} = (-a_{11}) \cdot (0) = 0$

Fig. 3.8

II: Model

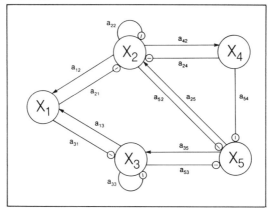

1. Path: $P_{51}^{(1)}$ Complementary Feedback

$P_{51}^{(1)} = 0$

Cannot get from X_1 to X_5 using only one variable. In fact,

$P_{ji}^{(1)} = 0$

for all cases when $j \neq i$

Not Relevant

2. Path: $P_{51}^{(2)}$ Complementary Feedback

$P_{51}^{(2)} = 0$

Not Relevant

3. Path: $P_{51}^{(3)}$ Complementary Feedback

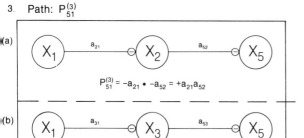
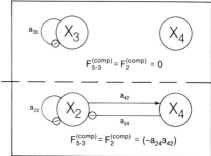

(a) $P_{51}^{(3)} = -a_{21} \cdot -a_{52} = +a_{21}a_{52}$ $F_{5-3}^{(comp)} = F_2^{(comp)} = 0$

(b) $P_{51}^{(3)} = -a_{31} \cdot -a_{53} = +a_{31}a_{53}$ $F_{5-3}^{(comp)} = F_2^{(comp)} = (-a_{24}a_{42})$

Fig. 3.8 (*continued*)

input to the system. This will be true mathematically even if the parameter is internal to the variable (for example, feeding rate). The feedback of the null complementary subsystem is $F_0^{(\text{comp})} \equiv -1$, just as $F_0 \equiv -1$.

Some examples of complementary subsystems and the associated open paths are illustrated in Figure 3.8.

EXAMPLE 3.3

Identify $p_{45}^{(5)}$ and give the complementary feedback for the system of Figure 3.8, model II.

Solution

Refer to the graph in Figure 3.9.

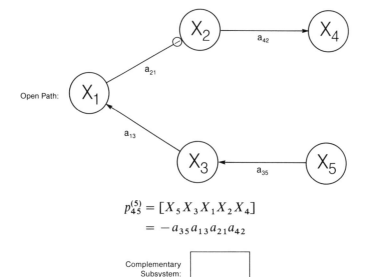

Fig. 3.9

$$p_{45}^{(5)} = [X_5 X_3 X_1 X_2 X_4]$$
$$= -a_{35} a_{13} a_{21} a_{42}$$

Complementary Subsystem:

Complementary Feedback: $F_{5-5}^{(\text{comp})} = F_0^{(\text{comp})} = -1$ □

Changes in Equilibrium Abundance

To determine the effect on the equilibrium values of all the variables if there is a change in a parameter of the growth rate function of a variable, ask these two questions:

1. Which parameters are affected? Identify the variables whose rate of change is altered by the parameter change (that is, the variables that have a direct functional dependence on the parameter).
2. What is the direction of the change in the growth rate? By "direction" we mean, Does the value of the growth rate increase or decrease with the new value of the parameter?

It is convenient to use mathematical symbols to express the answer to the above two questions. (It will be clear later and in the Appendix why the notation makes sense.) The symbols we use are

$$\frac{\partial f_i}{\partial c}$$

where ∂, the greek lowercase delta, denotes a change; f_i is the function for the growth rate of the ith variable, X_i (we do not have to know f_i); and c is one of the parameters which determines what the growth rate will be.

This set of symbols indicates we are looking at the immediate change in the growth rate function of variable i whenever parameter c changes. We can designate conveniently that the growth rate function is increasing, decreasing, or not changing by

$$\frac{\partial f_i}{\partial c} > 0 \quad \text{or} \quad \frac{\partial f_i}{\partial c} < 0 \quad \text{or} \quad \frac{\partial f_i}{\partial c} = 0$$

or equivalently as

$$\frac{\partial f_i}{\partial c} = (+) \quad \text{or} \quad \frac{\partial f_i}{\partial c} = (-) \quad \text{and} \quad \frac{\partial f_i}{\partial c} = 0$$

In analogous notation, the change in the equilibrium value of variable X_j due to a change in parameter c in one or more variables in the system network is denoted by $(\partial X_j^*/\partial c)$, where * identifies equilibrium value. The calculation of the change in variable X_j due to a change in parameter c is given by the formula (see Appendix)

$$\frac{\partial X_j^*}{\partial c} = \frac{\sum_{i,k} \left(\frac{\partial f_i}{\partial c}\right)(p_{ji}^{(k)})(F_{n-k}^{(\text{comp})})}{F_n} \tag{3.1}$$

The formula reads, take all the functions that include the parameter being changed $(\partial f_i/\partial c)$, trace each possible path $(p_{ji}^{(k)})$ to the jth variable (X_j) whose equilibrium value is being calculated, multiply each path by the appropriate complementary feedback $(F_{n-k}^{(\text{comp})})$. Sum $\sum_{i,k}$ for all functions and paths, and divide it by the overall feedback of the entire system of n variables (F_n). (For those unfamiliar with summation over two or more indexes, see the appendix.)

The formula of Equation 3.1 is given pictorial significance by the next sequence of illustrations, starting with the hypothetical three-variable model of Figure 3.10.

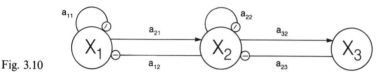

Fig. 3.10

A parameter change that decreases the growth rate of X_2 and has no effect on the growth rate of other variables is a negative parameter change to X_2, as shown in Figure 3.11.

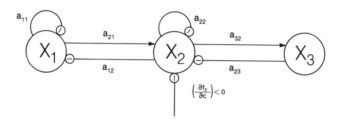

Fig. 3.11

The effect on the equilibrium level of X_3, $(\partial X_3/\partial c)$, via the only path from X_2 to X_3, $p_{32}^{(2)} = a_{32}$, is shown in Figure 3.12.

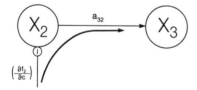

Fig. 3.12

This path has the complementary feedback shown in Figure 3.13.

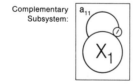

Fig. 3.13

The overall system feedback is $F_3 < 0$. The complementary feedback is $F_{3-2}^{(comp)} = F_1^{(comp)} = -a_{11} < 0$. Thus the change in the equilibrium abundance of X_3 due to a decrease in the growth rate of X_2 is calculated as

$$\frac{\partial X_3}{\partial c} = \frac{\left(\frac{\partial f_2}{\partial c}\right)(p_{32}^{(2)})(F_1^{(comp)})}{F_3}$$

$$= \frac{(-)(+)(-)}{(-)} = (-)$$

The equilibrium level of variable X_3 will decrease.

Again using the same three-variable system above, suppose outside events change parameters in more than one variable. Assume the parameter values change so as to decrease the growth rate of X_3 and increase that of X_2. This is likely to be a common result of a change in parameters because the function describing the feeding rate of X_3 on X_2 appears in the equation for the growth rates of both X_2 and X_3, but is positive for X_3 and negative for X_2. This phenomenon is shown in Figure 3.14.*

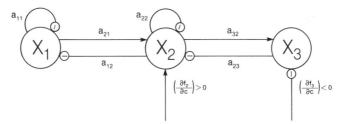

Fig. 3.14

How can we determine the change in equilibrium abundance of $X_1: \partial X_1/\partial c = ?$ First, take the effect of $\partial f_2/\partial c$, →(X_2) a parameter change that increases the growth rate function of X_2. The open path to X_1 is $p_{12}^{(2)} = -a_{12} < 0$, as is shown in Figure 3.15. The complementary subsystem is shown in Figure 3.16.

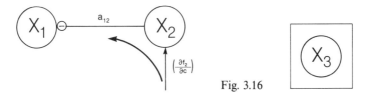

Fig. 3.15   Fig. 3.16

* We use the same symbol c to represent any parameter. This minimizes the number of symbols and subscripts used. An external event which causes parameter changes, however, might, and probably does, change different parameters in X_2 from those in X_3.

$$F_{3-2}^{(comp)} = F^{(comp)} = 0$$

The effect on X_1 from a positive parameter change to X_2 is

$$\frac{\partial X_1}{\partial c} = \frac{\left(\dfrac{\partial f_2}{\partial c}\right)(p_{12}^{(2)})(F_1^{(comp)})}{F_3} = \frac{(+)(-)(0)}{(-)} = 0$$

A change in X_1 due to the negative parameter change on the growth rate of X_3 is calculated in the same manner as above. The open path to X_1 is $p_{13}^{(3)} = (a_{23}a_{12}) > 0$, as illustrated in Figure 3.17.

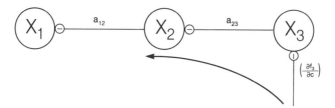

Fig. 3.17

The complementary subsystem is the null subsystem.

$$F_{3-3}^{(comp)} = F_0^{(comp)} = -1$$

then

$$\frac{\partial X_1}{\partial c} = \frac{\left(\dfrac{\partial f_3}{\partial c}\right)(p_{13}^{(3)})(F_0^{(comp)})}{F_3} = \frac{(-)(+)(-1)}{(-)} = (-)$$

The total effect on the equilibrium abundance of X_1 due to a change in parameters affecting variables X_2 and X_3 is

$$\frac{\partial X_1}{\partial c} = \frac{\left(\dfrac{\partial f_2}{\partial c}\right)(p_{12}^{(2)})(F_1^{(comp)}) + \left(\dfrac{\partial f_3}{\partial c}\right)(p_{13}^{(3)})(F_0^{(comp)})}{F_3}$$

$$\frac{\partial X_1}{\partial c} = \frac{(-)(-)(0) + (-)(+)(-)}{(-)}$$

$$= \frac{(0) + (+)}{(-)} = (-)$$

The parameter changes to X_2 and X_3 result in X_1 decreasing in magnitude. The positive parameter change to X_2 did not contribute to the decrease of X_1 because of the zero complementary feedback. We see that it is not enough to know the pathways from parameter changes to the variable of interest; you also must know about the interconnections (feedback) of the subsystem not included in the pathway.

Tables of Predictions

For any loop model with n variables there are n points of entry for parameter changes, one for each variable. A table of predictions can be constructed to show how each variable level will change, whether it increases, decreases, or remains the same, due to a parameter change to itself or any other variable. The formula of Equation 3.1 is applied n times.

Table 3.1 outlines the procedure for the simple three-variable linear system of Figure 3.18. The typical starting table setup is shown in Table 3.1.

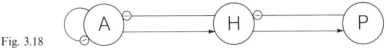

Fig. 3.18

Table 3.1 Table of predictions for Figure 3.17

Increase in rate of change of—	Effect of parameter change on equilibrium of—		
	A	H	P
A	.	.	.
H	.	.	.
P	.	.	.

A positive parameter change to A can possibly effect A, H, and P equilibrium levels. To predict what the direction of change will be we apply the formula of Equation 3.1, repeated below.

$$\frac{\partial X^*_j}{\partial c} = \frac{\sum_{i,k}\left(\frac{\partial f_i}{\partial c}\right)(p_{ji}^{(k)})(F_{n-k}^{(comp)})}{F_n} \quad (3.2)$$

In applying the formula we should note that F_n in this case is F_3 and $F_3 < 0$. This will remain the same for all the calculations. The overall system feedback, F_n, is only calculated once in order to fill in the table of predictions.

The $\sum_{i,j}$ is for $i = A, H, P$ and $j = A, H, P$, with j changing most rapidly. Take each parameter change (each i) in turn and for that parameter change calculate its effect on each variable (each j). By calculating only one row at a time we effectively fix i at one of its values. An alternative way to view this is

Table 3.2 Calculations necessary for constructing a table of predictions

Parameter change	Growth rate change*	Path end point	Path length	Path value	Complementary subsystem	$F_{3-k}^{(-)}$	Effect
$i = A$	$\dfrac{\partial f_A}{\partial c} = (+)$	$j = A$	$k = 1$	$p_{AA}^{(1)} = 1$	$[HP]$	$F_2^{(-)} = (-)$	$\dfrac{\partial A}{\partial c} = \dfrac{(+)(1)(-)}{(-)} = (+)$
			$k = 2$	$p_{AA}^{(2)} = 0$		By definition there are no open paths from A to itself of length 2 or greater.	
			$k = 3$	$p_{AA}^{(3)} = 0$			
		$j = H$	$k = 1$	$p_{HA}^{(1)} = 0$		No open paths of length 1 from A to H.	
			$k = 2$	$p_{HA}^{(2)} = (+)$	$[P]$	$F_1^{(-)} = 0$	$\dfrac{\partial H}{\partial c} = \dfrac{(+)(+)(0)}{(-)} = 0$
			$k = 3$	$p_{HA}^{(3)} = 0$		No open paths of length 3 from A to H.	
		$j = P$	$k = 1$	$p_{PA}^{(1)} = 0$		No open paths of lengths less than 3 from A to P	
			$k = 2$	$p_{PA}^{(2)} =$			
			$k = 3$	$p_{PA}^{(3)}$	$[\text{null}]$	$F_0^{(-)} = -1$	$\dfrac{\partial P}{\partial c} = \dfrac{(+)(+)(-)}{(-)} = (+)$

$i = H$ 　　 $j = A$ 　　 $k = 1$ 　　 $p_{AH}^{(1)} = 0$ 　　　　　　　　　　　　 No open path.

$\boxed{\dfrac{\partial f_H}{\partial c} = (+)}$

$k = 2$ 　　 $p_{AH}^{(2)} = (-)$ 　　 [P] 　　 $F_1^{(-)} = 0$ 　　 $\boxed{\dfrac{\partial A}{\partial c} = \dfrac{(+)(-)(0)}{(-)} = 0}$

$k = 3$ 　　 $p_{AH}^{(3)} = 0$ 　　　　　　　　　　　　 No open path

$j = H$ 　　 $k = 1$ 　　 $p_{HH}^{(1)} = 1$ 　　 [A][P] 　　 $F_2^{(-)} = 0$ 　　 $\boxed{\dfrac{\partial H}{\partial c} = \dfrac{(+)(1)(0)}{(-)} = 0}$

$k = 2$ 　　 $p_{HH}^{(2)} = 0$ 　　　　　　　　　　　　 No open path

$k = 3$ 　　 $p_{HH}^{(3)} = 0$ 　　　　　　　　　　　　 No open path

$j = P$ 　　 $k = 1$ 　　 $p_{PH}^{(1)} = 0$ 　　　　　　　　　　　　 No open path

$k = 2$ 　　 $p_{PH}^{(2)} = (+)$ 　　 [A] 　　 $F_1^{(-)} = (-)$ 　　 $\boxed{\dfrac{\partial P}{\partial c} = \dfrac{(+)(+)(-)}{(-)} = (+)}$

$k = 3$ 　　 $p_{PH}^{(3)} = 0$ 　　　　　　　　　　　　 No open path

(continued overpage)

Table 3.2 Continued.

Parameter change	Growth rate change*	Path end point	Path length	Path value	Complementary subsystem	$F^{(-)}_{3-k}$	Effect
$i = P$	$\dfrac{\partial f_P}{\partial c} = (+)$	$j = A$	$k = 1$	$P^{(1)}_{AP} = 0$			No open path
			$k = 2$	$P^{(2)}_{AP} = 0$			No open path
			$k = 3$	$p^{(3)}_{AP} = (=)$	[null]	$F^{(-)}_0 = -1$	$\dfrac{\partial A}{\partial c} = \dfrac{(+)(+)(-1)}{(-)} = (+)$
		$j = H$	$k = 1$	$p^{(1)}_{HP} = 0$			No open path
			$k = 2$	$p^{(2)}_{HP} = (-)$	[A]	$F^{(-)}_1 = (-)$	$\dfrac{\partial H}{\partial c} = \dfrac{(+)(-)(-)}{(-)} = (-)$
			$k = 3$	$p^{(3)}_{PP} = 0$			No open path
		$j = P$	$k = 1$	$p^{(1)}_{PP} = 1$	[AH]	$F^{(-)}_2 = (-)$	$\dfrac{\partial P}{\partial c} = \dfrac{(+)(1)(-)}{(-)} = (+)$
			$k = 2$	$p^{(2)}_{PP} = 0$			No open path
			$k = 3$	$p^{(3)}_{PP} = 0$			No open path

that for each increment in *i*, that is, as *i* goes from A to H to P, we are moving down the rows; while for each increment in *j*, as *j* goes from A to H to P, we are moving across the columns.

The details of the calculations are given in Table 3.2, and the predictions contained in Table 3.3.

Table 3.3 Table of predictions

Increase in rate of change of—	Effect of parameter change on equilibrium abundance of—		
	A	H	P
A	+	0	+
H	0	0	+
P	+	−	+

Consider the following simple model. We will use this model to construct a complete table of predictions.

Our model is of a lake in which nitrate (N) and phosphate (P) are consumed by green algae (G). Blue-green algae (B) only consume phosphate, fix nitrogen, and inhibit the growth of the green algae. The signed digraph for this system is shown in Figure 3.19.

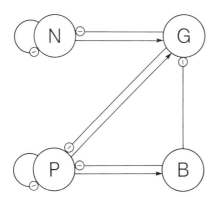

Fig. 3.19

We can begin to construct a table of predictions by asking what happens to the equilibrium levels of each variable if there is a change in the environment which causes a positive parameter change shown in Figure 3.20*. Table 3.4 outlines this set of predictions.

---

* By "positive parameter change" we mean that a parameter has changed, either by increasing or decreasing in value, resulting in an increased growth rate of the variable.

Once the table is calculated for a positive parameter change, the predictions for a change in parameters that cause the growth rate of the variables to decline ("negative parameter change") is obtained by reversing all the signs of the table.

Fig. 3.20

Table 3.4 Table of predictions for Figure 3.20

Positive parameter input to—	Effect of change on equilibrium abundance of—			
	N	P	G	B
→N	−	0	+	−
P				
G				
B				

Now let the change increase the rate of growth of the phosphate, illustrated in Figure 3.21. Table 3.5 charts the predictions for this model.

Fig. 3.21

Table 3.5 Table of predictions for Figure 3.21

Positive parameter change to—	Effect of change on equilibrium abundance of—			
	N	P	G	B
N				
→P	+	0	?	+
G				
B				

EXAMPLE 3.4

Complete the table of predictions for the lake model of Figure 3.19.

Solution

This completes the mechanics of loop analysis. The use of qualitative models, however, just begins with the mechanics. The next two chapters demonstrate some of the kinds of thinking that go beyond the signed digraph algorithms.

The examples we presented in this chapter illustrate the properties of qualitative analysis that are not obvious. A variable may change in the opposite direction from what common sense would lead us to expect (N in Table 3.6). A variable may be linked to others, may change other variables,

Table 3.6 Table of predictions for Figure 3.19

Positive parameter input to—	Effect of change on equilibrium abundance of—			
	N	P	G	B
→ N	−	0	+	−
→ P	+	0	?	+
→ G	−	0	+	−
→ B	+	?	?	+

yet not change itself. This does not imply any insensitivity to the direct impact of its neighbors or the environment—only that the impact is passed along to another variable. One final note: the effect of an environment input may not be observable at its point of entry, but only in variables several steps away.

□

Links from Equations

Armed with the notation of a partial derivative, which is described mathematically in the Appendix, we can now give a more formal definition of a link. In loop models the links between variables are determined by the first partial derivatives of the growth equations evaluated at equilibrium. Simply put, in a system of n variables each variable is represented by a differential equation that gives the rate of change of the abundance or level of each variable according to a general formula, which for the ith variable is

$$\frac{dX_i}{dt} = f_i(X_1, X_2, X_3, \cdots, X_n; c_1, c_2, c_3, \cdots)$$

In words, the growth rate of X_i is a function, f_i, of the levels of X_i for some or all of the other variables in the system, and usually itself, and a set of parameters, c. The parameters in ecological systems represent aspects of the general environment, typically physical characteristics like precipitation and temperature, and biological properties like birth and death rate constants or feeding efficiency

The links are obtained in the formal sense by taking first partial derivatives of the growth rate function,

$$a_{ij} \equiv \frac{\partial}{\partial X_j}\left(\frac{dX_i}{dt}\right)_{X=X^*}$$

$$= \frac{\partial}{\partial X_j}\left(f_i(X_1, X_2, X_3, \cdots, X_n; c_1, c_2, c_3, \cdots)\right)_{X=X^*}$$

where j takes on the values of 1 through n. (All variables are included, whether or not they explicitly appear in the function f_i.)

EXAMPLE 3.5

Determine the links between the variables H and N of Example 3.1 from Equations E3.1.1 and E3.1.2.

Solution

$$\frac{\partial}{\partial N}\left(\frac{dN}{dt}\right)_{N=N^*, H=H^*} = a_{NN} = -pH|_{N^*, H^*} = -p\left(\frac{Ir}{p\theta}\right) = \frac{-Ir}{\theta}$$

$$\frac{\partial}{\partial H}\left(\frac{dN}{dt}\right)_{N=N^*, H=H^*} = a_{NH} = -pN|_{N=N^*, H=H^*} = \frac{-p\theta}{r}$$

$$\frac{\partial}{\partial N}\left(\frac{dH}{dt}\right)_{N=N^*, H=H^*} = a_{HN} = rH|_{N=N^*, H=H^*} = r\left(\frac{Ir}{p\theta}\right) = Ir$$

$$\frac{\partial}{\partial H}\left(\frac{dH}{dt}\right)_{N=N^*, H=H^*} = a_{HH} = (rN - \theta)|_{N=N^*, H=H^*} = 0$$

□

4 Qualitative Predictions

In this chapter we extend the techniques of the previous chapter to predictions of turnover rates and correlations among variables, and then apply them to the behavior of systems whose graphs are presumed known.

The graph may be constructed from knowledge of the interactions of variables in a particular system. It might be set up to examine a general conceptual issue: under what circumstances might pesticides increase pest problems? Can natural selection cause a species to become less abundant? How do specialist and generalist competitors interact? How does the attempt to manage a system affect its behavior? The graph also may be produced in an attempt to account for a set of observations. In any case, we start by taking the graph as given.

The models used in the examples presented in this chapter are suggested by real systems, but have been simplified to remove ambiguities for pedagogical purposes. We will consider various kinds of ambiguities later in this book.

In this chapter we introduce differential equations and the notation of differentiation and partial derivatives. Readers who are unfamiliar with this either should skip over the detail in the following sections, until Chapter 6, or examine an elementary textbook such as Hildebrand (1962).

Equilibrium Levels and Turnover Rates

In previous examples we examined changes in the equilibrium levels of the variables which resulted from parameter change. In some cases we concluded that despite a change of input a particular variable did not change. What was meant, of course, was only that the equilibrium level (the number of organisms in a population, the concentration of a substance) did not change. Other changes, however, may take place in those populations. In particular, the turnover rate of a variable (the death rate of an organism, the residence time of a chemical) can change drastically even if the equilibrium level does not. Therefore, predictions about turnover rate allow for an additional set of tests of the model and provide new insight into its dynamics.

Turnover rates are defined in our ecological examples as the reciprocal of life expectancy of an organism or as the residence time of a chemical.

In a system composed of a nutrient (N), a consumer (C), and a predator (P) of the consumer, an increased input to the nutrient will increase the nutrient concentration and the predator numbers but not the consumer population, as shown in Figure 4.1.

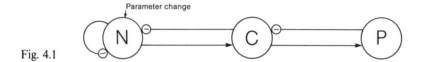

Fig. 4.1

There is greater "flowthrough," however. The level of nutrient is increased, consumer animals are better nourished and more are produced. Since there are more predators, however, more of the consumers are being eaten. The higher birth and death rate of the consumers shifts the age distribution toward the younger age classes. This shift may be accompanied by a smaller body size or even by a different proportion of animals in different development stages (for example, the proportion of immature to mature; or, of different instars, if they are larvae; or, of larvae-to-adult ratio). We can infer the change in age structure of the population by examining birth and death rates without having direct observation of age.

Suppose in the model of Figure 4.1 the input came into the system by way of the consumer—for example, a decrease in temperature reduced heat stress mortality. This would have no effect on the nutrient or consumer levels, as long as predators increase. The cause of death in the consumer but not the death rate has changed. Since the concentration of nutrients is unchanged the reproductive rate of the consumer is not affected via the nutrients. Of course there may be some change in the consumer reproductive rate, for reasons other than nutrient increase.

For example, if the parameter change is a greater heat tolerance in the consumer, mortality is reduced due to this cause. Mortality due to predation would increase due to an increase in predators. Under these assumptions, and without any change in food supply, the birth rate is unchanged and the population level is unchanged, so that mortality is also unchanged. Therefore the increase due to predation exactly balances the greater survival due to heat tolerance. Then the age distribution is unaltered.

The predator population increases until it is just able to hold the consumer population down to its previous level. Since predator nutrition and mortality are unaltered, its age distribution is also unchanged.

Suppose instead that the parameter change in the consumer acts by reducing fecundity. A decrease in the birth rate reduces the production of food for the predator whose population declines until the consumer population readjusts upward to its previous level, just enough to sustain the predators. The consumer now has reduced birth and death rates; therefore, the population is older.

EXAMPLE 4.1

What happens to the nutrient turnover in each example of parameter change of the model in Figure 4.1?

Solution

Input to nutrient leaves the consumer level unchanged. Therefore the rate of consumption, and hence the residence time of the nutrient, remain unaltered. Similarly, input to the consumer leaves the consumer level unchanged, and therefore the nutrient turnover rate does not change. Finally, an input to the predator reduces the consumer, increasing both the nutrient level and the turnover time.

We now can expand the table of predictions by giving the change in turnover rate in parentheses next to the predicted change in equilibrium level, as shown in the table below:

	N	C	P
N	+(0)	0(+)	+(0)
C	0(0)	0(0)	+*(0)
P	+(−)	−(+)	+*(+)

* The input is assumed to act by reducing mortality.

More practice in determining turnover rates can be had from Example 4.4 and the loop diagrams of Figure 4.10. □

Correlations from Predictions

In laboratory experiments, or in field situations where we know the input (say an increase in fishing acting as a negative input to P in the model of Figure 4.1), we can produce a set of predictions about the direction of change of each variable. But it is often the case that we do not know where the input enters the system or whether it is positive or negative. It is still possible to make predictions from the model and to use them to identify the source of input by examining the correlations among variables either from site to site or over time periods far enough apart so that the moving equilibrium assumption is valid.

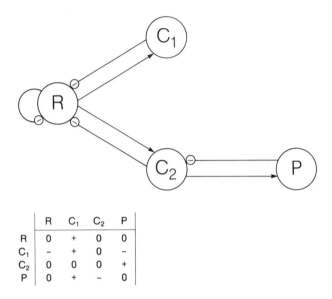

Fig. 4.2

	R	C_1	C_2	P
R	0	+	0	0
C_1	−	+	0	−
C_2	0	0	0	+
P	0	+	−	0

Consider Figure 4.2. The table of predictions for the model presented therein is shown at the bottom of the figure. The procedure is as follows: If two variables change in the same direction in response to a particular input, then they will show positive correlation over a range of that parameter. If they change in opposite directions, they will show a negative correlation. If one or both show zero response, then this input would result in no correlation between them.

From the table we see that input to R generates no correlation among variables since only C_1 changes. With input to C_1, both R and C_1 change in opposite directions, R and P in the same direction, and C_1 and P in opposite directions. Thus we can construct a set of four tables, giving the correlation pattern for the variables in response to each of the inputs.

The diagonal element is the correlation of a variable with itself, and therefore positive (in fact, +1) if there is any variation at all; otherwise zero. (See below for further discussion of zeros.) Now look at Table 4.1. There you see that each part of the table has ten entries, including diagonals. The predictions from different inputs coincide for some entries and differ for others. Excluding diagonals, inputs to R and C_1 agree in three predictions

QUALITATIVE PREDICTIONS

Table 4.1 Correlations among variables with parameter changes (input) to each separately

Input to:	R				C_1				C_2				P			
	R	C_1	C_2	P	R	C_1	C_2	P	R	C_1	C_2	P	R	C_1	C_2	P
R	0	0	0	0	1	−	0	+	0	0	0	0	0	0	0	0
C_1		1	0	0		1	0	−		0	0	0		1	−	0
C_2			0	0			0	0			0	0			1	0
P				0				1				1				0

and differ in three. R and C_2 agree in all predictions, and R and P differ in one. C_1 and C_2 differ in five, C_1 and P differ in four, and C_2 and P differ in one prediction. Therefore by examining the correlations we could identify the source of input.

In real life, no correlation is likely to be exactly zero. Sampling error will show differences even when none are expected on the average. Therefore we have to reinterpret the table so that a 1 indicates a large variance and a 0 a very small one on the diagonal. Zero correlations in the positions off the diagonal would really be nonsignificant statistically.

Suppose that the real pattern of correlations does not correspond to any of the tables. We then have several possible choices to make. One possibility is that there is more than one major input. For example, suppose that there is a positive correlation between C_1 and C_2, an outcome that does not correspond to any input. But if there are inputs both to R and to P, and these are negatively correlated, then when C_2 decreases due to a positive input to P, C_1 will increase for the same reason but decrease due to negative input to R. If the negative input to R is strong enough, C_1 will show a net decrease. Conversely, negative inputs to P will be associated with positive input to R and again C_1 and C_2 can change in the same direction and be positively correlated.

Since there are six correlations among variables and six possible correlations among inputs, it is always possible to invent a correlation pattern among inputs which would fit any observed pattern among variables. Therefore this procedure is not sufficient to establish the sources of input. All it can do is argue, if the observed pattern is due to this graph, that it must come from multiple inputs correlated in a particular way. We would then have to demonstrate that the inputs indeed have that pattern.

There are three possible states $(-, 0, +)$ for each entry in the correlation table, and therefore 3^6 or 729 possible tables, of which only four correspond to single inputs. Therefore if a single input is consistent with the observations there is a strong presumption that our interpretation is not accidental. Choosing two inputs would give us six possible pairs of variables and three correlation states for each pair of $(-,0,+)$ so that we could produce eighteen of our 729 possible tables with paired inputs. Therefore the chance of explaining the table with two inputs is $18/729 = 0.025$. Beyond two inputs it is increasingly easy to find a hypothesis to fit the data.

A second alternative is that the graph leaves out some important links, includes links that are really absent, or assigns the wrong sign. The appropriate procedure is to examine a whole set of alternative models. Finally, it may be that the loop analysis assumption is inapplicable. That is, the changes may occur too rapidly for the assumption of moving equilibrium and we have to move on to time-averaging methods (see Chapter 7).

The statistical testing of a hypothesis about the signed digraph requires a comparison of the observations with predictions and the calculation of the probability of getting at least as good a fit if the model is incorrect. That probability depends of course on what we assume the "true" state of nature to be, the null hypothesis. The choice of a null hypothesis is not a trivial question. We might ask, suppose the correlations among variables are random numbers drawn from a uniform distribution from -1 to $+1$. Then $+$ and $-$ are equally probable observations. Each of the possible $2^{n(n-1)/2}$ tables is then equally likely, and the probability of predicting all the signs of the correlations is $2^{-n(n-1)/2}$. We also could allow that there are three alternatives: $+, 0,$ or $-$. The trouble is there usually will not be any zero correlations observed. Therefore we have to decide that correlations between some $-r_1$ and $+r_1$ will be treated as if they were zero. We might choose r_1 as the level of statistical significance for our sample size. Another possibility would be to observe the actual numbers of $+, -$ (and effectively zero) correlations, and consider what the probability of a given fit would be if these $+, -,$ and 0 were distributed at random. Finally, we might compare our table of predictions to one or more alternative graphs and determine what set of graphs is compatible with the observations.

Consider the hypothetical table of predictions, Table 4.2 (the associated digraph is not shown).

Table 4.2 Hypothetical table of predictions

	X_1	X_2	X_3	X_4	X_5
X_1	+	+	−	0	+
X_2	−	−	0	−	−
X_3	0	+	+	0	0
X_4	−	+	0	+	+
X_5	+	+	+	0	+

A pairwise comparison of columns may in some cases reveal a correlation pattern. Of the ten possible comparisons, we illustrate five in Table 4.3.

Table 4.3 Five possible comparisons

	X_1	X_2	X_1	X_3	X_1	X_5	X_3	X_4	X_4	X_5
X_1	+	+	+	−	+	+	−	0	0	+
X_2	−	−	−	0	−	−	0	−	−	−
X_3	0	+	0	+	0	0	+	0	0	0
X_4	−	+	−	0	−	+	0	+	+	+
X_5	+	+	+	+	+	+	+	0	0	+

In Table 4.3 we have assumed that the data collection is across a wide geographic range and therefore each variable has at some place encountered a change in one or more of its parameters.

The correlation between X_4 and X_5 will be positive if parameter change enters the system at X_2 or X_4, and will be zero otherwise. Since X_5 changes whenever X_4 does but also when X_4 does not, we expect roughly that X_5 is more variable than X_4 and the two show a possible weak correlation. The weaker the correlation the less variable X_4 is compared to X_5. Similarly, we expect there to be no correlation between X_3 and X_4 if a single node accounts for most of the parameter variation at any one location. But if parameters entering at X_4 and X_5 are positively correlated, then X_3 and X_4 may show positive correlation.

The interpretation of the correlations among the other variables is more complex. The correlations of X_1 and X_2 will have the same sign as the correlation of X_1 and X_5 except if the parameter enters at X_5. In that case, opposite sign correlations would be associated with a positive correlation of X_1 and X_3.

From Model to Predictions

We now will consider applications of loop models.

It frequently has occurred that the application of pesticides has resulted in increased pest problems (deBach, 1971). One reason for this is that pesticides may kill not only the target pest but also its predators or parasitoids (other insects, usually wasps or flies, that lay their eggs in the bodies of their hosts). This is the traditional explanation or folk explanation. It is not obvious why, if both predator and prey are killed, the prey (the pest species) may increase sometimes. The models shown below suggest why this happens and also circumstances in which it may not happen.

Suppose that a pesticide kills only the herbivore. From the middle row of the table of predictions for the first model (Fig. 4.3) we see that the equilibrium plant and herbivore populations would not change, while the carnivore, C, would decrease. (Recall the table is calculated under the assumption that a change in parameters results in an increase in the growth rate. In this example a pesticide is increasing the mortality rate parameter of the herbivore; therefore, the + sign in the second row is reversed.)

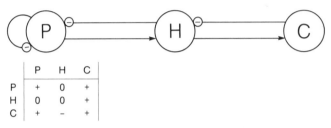

	P	H	C
P	+	0	+
H	0	0	+
C	+	−	+

Fig. 4.3 P = plant biomass, H = herbivorous insect (the pest), C = carnivore (predator or parasitoid).

The decrease in the carnivore C is wholly the result of the pesticide killing the herbivore H, thus reducing its food supply. The reduction in C results in fewer H being eaten, compensating for the increased mortality due to the pesticide. This is a very different explanation than the one usually offered.

Now suppose that the pesticide affects both insect species. Then the impact would be read from the table as negative inputs both to H and C. The negative input to H leaves H unchanged but reduces C. The negative input to C not only reduces C but also increases H. The final result is that carnivores decrease more than enough to compensate for the pesticide-induced mortality of the herbivore, and the herbivore increases, thus reducing the plant biomass.

It is important to note that this result does not presume that predators are more sensitive to pesticides than their prey. Rather it depends on the structure of the system. The direct effect of the pesticide on the herbivore is offset by the indirect effect on the carnivore. But for the carnivore, the direct toxic effect, and the indirect effect, the poisoning of some of its prey, both act in the same direction.

This situation would be rather different if the carnivores were not a satellite variable. Suppose instead that C has an alternative prey that feeds on plants in other habitats that are inaccessible to the pesticide, as depicted in Figure 4.4.

A negative input to H and C still reduces C, but the effect on H is ambiguous: The change in H is the direct toxic impact on H times the feedback of the complementary subsystem $[CH_2P_2]$, the direct toxic impact

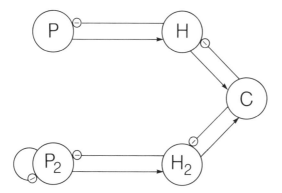

Fig. 4.4

on C times the negative link between C and H, a_{HC}, times its complementary feedback $[P_2 H_2]$. Depending on these other terms, H may either increase or decrease.

The carnivore itself may be preyed upon by its own specialized predator or parasitoid C_2, as shown in Figure 4.5. The table of predictions for the two levels of predators model is outlined in the table at the bottom of the figure. As we see in this figure, and from the table of predictions, a pesticide killing H, C, and C_2 reduces the parasitoid C_2 and protects the crop.

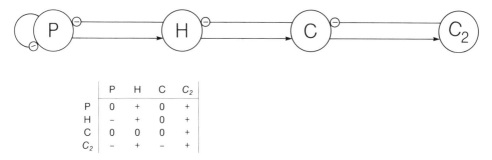

	P	H	C	C_2
P	0	+	0	+
H	−	+	0	+
C	0	0	0	+
C_2	−	+	−	+

Fig. 4.5

Predator–prey relations are not the only pathways that can thwart a pesticide program. Suppose that two herbivores utilize the crop, while one of them has an alternative host. This is the model of Figure 4.6, where P_1 is the common crop plant for both herbivores, R_1 and R_2, while P_2 is a plant eaten only by the second herbivore. A pesticide which diminishes H_2 will have no effect on the crop biomass because the alternative herbivore H_1 benefits from the reduction in H_2.

The loop analysis shown in Figure 4.6 does not explain why any particular program failed. Rather, it shows how this can happen in some community

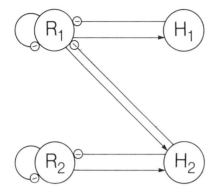

Fig. 4.6

structure, suggests where to look for an explanation, and suggests that the community structure must be analyzed before undertaking pesticide application.

In another study of pesticide application, we consider models which incorporate mosquito life-history. Mosquitos have a life cycle from egg to larva to pupa in water, after which the adult emerges and is a pest and sometimes a vector. Since we are interested in affecting the abundance of adult mosquitos by interventions at any one stage, our model divides the population of mosquitos into life-stage classes (Fig. 4.7).

The life-stages are self-damped because they are no longer self-reproducing variables. For example, suppose

$$\frac{dE}{dt} = fA - (m + d)E$$

where f is the rate of egg production by adults and m and d are the rates of removal of eggs from the community by maturing into larvae (m) or dying (d). The coefficients f, m, and d may depend on other variables in the system but not on E. If we use a prime to denote differentiation with respect to time, that is, $E' = dE/dt$, then the derivative of E' with respect to E yields

$$\frac{\partial E'}{\partial E} = -(m + d)$$

so that the self-damping of eggs is its turnover rate, an inverse measure of the duration of the egg stage.

EXAMPLE 4.2

Show that the adults, larvae, and pupae mosquito variables of the model in Figure 4.7 are self-damped. (Note: you may wish to skip this until you read Chapter 6.)

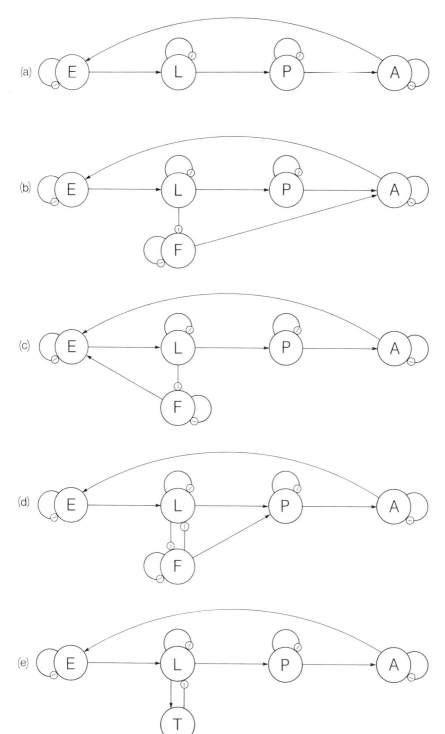

Fig. 4.7

Solution

The growth rate of the population of mosquito adults is

$$\frac{dA}{dt} = f_P P - d_A A$$

where f_P is the rate of maturation of pupae into adults and d_A is the death rate of adults. Then

$$\alpha_{AA} = \frac{\partial A'}{\partial A} = -d_A$$

The growth rate of the larvae population is

$$\frac{dL}{dt} = f_E E - (m_L + d_L)L$$

with f_E the rate of maturation of eggs into larvae, m_L maturation of larvae into pupae, and d_L the death rate of larvae. Then

$$\alpha_{LL} = \frac{\partial L'}{\partial L} = -(m_L + d_L)$$

The growth rate of the pupae population is

$$\frac{dP}{dt} = f_L L - (m_P + d_P)P$$

where f_L is the maturation of larvae into pupae and m_P and d_P are the rates of maturation and death of pupae, respectively. Then

$$\alpha_{PP} = \frac{\partial P'}{\partial P} = -(m_P + d_P) \qquad \square$$

The ecology of the mosquito may involve its own food supply and possible predators. In Figure 4.7b,c,d we have larvae feeding on some food supply, F, which is also self-damped. Food can result in healthier adults which live longer and therefore have a greater capacity to spread disease (Fig. 4.7b), or food may result in increased fecundity (Fig. 4.7c) and therefore increase egg production. It has been found by Frogner (1975) that for tree-hole mosquitos increased food supply speeds up development, transforming larvae into pupae more rapidly (Fig. 4.7d). Finally, in Figure 4.7e we consider the effect of a predator on the larvae such as the predaceous larvae of the mosquito *Toxorhynchites*.

Figure 4.7a shows a simple life cycle. There are positive and negative feedbacks, so that this system seems to have undetermined stability. This is neutral stability because feedback at Level 4, F_4, is 0. Despite the self-damping of each life-stage separately, the positive loop of length 4 exactly balances the product of the self-damping loops so that mosquitos as a whole form a single self-producing and unself-damped variable in this case. Therefore, the table of predictions is not shown. The table of predictions for the models of Figure 4.7b–e are given in Table 4.4.

Table 4.4 Predictions for models of Figure 4.7b–e

	(b)						(c)				
	E	L	P	A	F		E	L	P	A	F
E	+	+	+	?	−	E	+	+	+	+	−
L	?	+	+	?	−	L	?	+	+	+	−
P	+	+	+	+	−	P	+	+	+	+	−
A	+	+	+	+	−	A	+	+	+	+	−
F	+	+	+	+	0	F	+	+	+	+	0

	(d)						(e)				
	E	L	P	A	F		E	L	P	A	T
E	?	+	?	?	−	E	+	0	0	0	+
L	?	+	?	?	−	L	0	0	0	0	+
P	?	+	?	?	−	P	+	0	+	+	+
A	?	+	?	?	−	A	+	0	0	+	+
F	?	−	?	?	−	T	−	−	−	−	+

The relationships shown in Table 4.4 contain many ambiguities which would be important for mosquito control, and which depend on the relative strength of effects. In Table 4.4b, an increase in egg or larval mortality may actually increase the adult population. Furthermore, regardless of whether A is increased or decreased, F acts on A by reducing mortality. This increases the average size of A, making it a more effective vector. The ambiguity in Table 4.4c does not have practical significance. The model of Figure 4.7d compounds the ambiguities. Now interference with any life cycle stages may increase the adult population. Finally, the model of Figure 4.7e shows that a specialized larval predator behaving as a satellite variable to larvae buffers the adult population from egg and larval mortality.

There are also ambiguities in the stability analysis. The negative loops passing through food in the model of Figure 4.7c and d may induce stable or unstable oscillations. We examine the consequences of ambiguity in models and predictions in the last section of this chapter.

EXAMPLE 4.3

In the three models shown in Figure 4.8 there is the potential for positive feedback at one or more levels. Identify the positive feedback and the consequences for the system behavior that is counterintuitive.

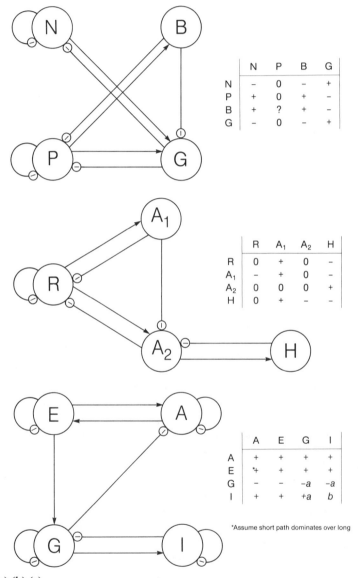

Fig. 4.8 (a) (b) (c)

Solution

(a) At Level 3 F_3 is ambiguous: negative feedbacks exist from the product of the disjunct loop [NG] with [P], from the product [PG] with [N], and from the product [PB] with [N]. Positive feedback comes from the length 3 loop [BGP]. In this model we assume the long loop feedback is stronger than the disjunct product of shorter loops. Therefore, when N is added to the system at an increasing rate, the N level actually decreases. This is a potential model of a lake system with nitrogen N, phosphorus P, green G, and blue-green (B) algae.

(b) In this system there is a resource, R, two consumers of the resource A_1 and A_2, and a consumer H of A_2 (we have shown a model similar to this before). Here the complement of H is the positive feedback loop [RA_1A_2]. Therefore a parameter change that improves life for H would reduce its numbers. The biological reason for this is that the initial increase in H reduces A_2, improving A_1's competitive position for R. The increase in A_1 reduces A_2 even further, so that A_2 is reduced more than it would have been by H alone. This overcompensation finally reduces H.

(c) This model is based on the regulation of blood sugar G by insulin, I, and epinephrine (adrenalin), E. A represents anxiety. Anxiety brings out adrenalin, which in turn is responsible for the symptoms of anxiety. A fall in glucose level induces anxiety (the hypoglycemic response) either by way of epinephrine or through some other pathway as shown in the figure, with intermediate variables omitted. The symbols a and b in the table are not known: a is the value of the feedback of the [EA] subsystem. If the product of the self-damping terms exceed the positive feedback loop then $a < 0$ and increased input of G increases G levels. But if $a_{EA}a_{AE} - a_{AA}a_{EE} > 0$, a is positive and we would observe the anomalous responses that increasing glucose intake reduces glucose and reduces insulin, and that insulin increases glucose levels. If this were actually observed, it would most likely be discarded as experimental error or attributed to some genetic anomaly in the insulin molecule. But it would really depend on the relative magnitudes of the physiological and psychological parameters of the [EA] subsystem. □

We often know pairwise relationships among factors in physical systems but the joint relation among all of them is not known. The factors contributing to the geometrical cross-section of a stream, which in part determine the stream discharge, are known but the exact nature of their interactions is not (Slingerland, 1981). At any location the volume of water per unit time transported by a river obviously depends on the flowing water velocity and the "width" and "depth" of the river. Simultaneously, stream discharge alters the alluvial channel configuration. Local conditions also contribute to the geometrical shape of a stream.

The local variables determining stream geometry properties have been simplified by Slingerland to include: R, the hydraulic radius; f, a friction factor; V, the mean stream velocity; Q, the stream bedload transport rate; and S, the stream bed slope. Simply put, the factors are the cross-sectional shape influence on water movement (R,f), speed of the water (V), speed of removal of bottom particles (Q), and how steep is the stream (S) (that is, it is on a flat or a hill). Each of these factors determines the hydraulic radius (R)—a geometric measure of the alluvial channel—but the *exact* functional relation among all six variables is not known, except for controlled estuary studies. The direction of interactions has been described (Slingerland, 1981), as illustrated in Figure 4.9.

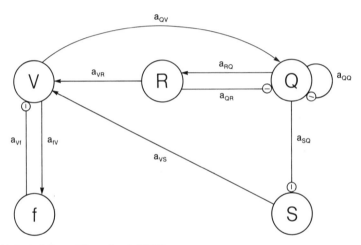

Fig. 4.9 (Adapted from Slingerland, 1981)

An increase in hydraulic radius (R) means the channel slope, maximum flow depth, and other geometric characteristics increase, which is assumed will increase stream velocity. Bedload discharge (Q), in the absence of changing conditions (for example, the stream going from laminar to turbulent flow), generally declines as the hydraulic radius (R) increases, simply because the deeper the channel the slower the removal of particles. Friction (f) retards the water velocity (V), while an increased water velocity amplifies the frictional drag. Clearly, a swift moving stream transports greater bedload (Q). The self-damping of bedload is because streams have a saturation level in their bedload capacity (Slingerland, 1981) since above a given bedload the river would become clogged to further flow—an event consistent with facts on a geological time scale but not in terms of a decade or so. Also, the transport of sediment into the area would also add to the self-damping. As the bottom material of the stream is carried downstream, including scour of the bed and banks, there is a reduction in the slope (S) of the river. Moreover, a decreasing slope decreases stream velocity.

Slingerland gives intuitive evidence for the relative strengths of the interactions among the variables. Friction factors can change in hours (or days), while changes in hydraulic radius take years. Slope alteration is on the order of decades. A notable exception to the above observations is for glacial outwash streams, the friction factor outpaced by changes in the hydraulic radius with an ensuing instability of braiding, or criss-crossing of the stream down the valley floor.

From the model it is obvious the first stability criterion is not met because $F_5 = 0$. Also, there is the potential for F_3 to be positive, since the loop of $[RVQ]$ is positive. As long as the self-damping on the bedload transport times the friction–velocity subsystem feedback plus the velocity–bedload–slope subsystem feedback are strong then this will not be the case. Symbolically, we have

$$|(a_{QQ})(a_{Vf}a_{fV}) + (a_{QV}a_{SQ}a_{VS})| > |(a_{QV}a_{RQ}a_{VR})|$$

The second criterion for stability is

$$F_1 F_2 F_3 + F_3^2 < F_4 F_1^2 - F_1 F_5 \tag{4.1}$$

where $F_1 = [-a_{QQ}]$
$F_2 = [-(a_{Vf}a_{fV}) - (a_{RQ}a_{QR})]$
$F_3 = [-(a_{QQ})(a_{Vf}a_{fV}) + (a_{QV}a_{RQ}a_{VR} - a_{QV}a_{SQ}a_{VS})]$
$ = [-(a_{QQ})(a_{Vf}a_{fV}) + d]$
$F_4 = [-(a_{Vf}a_{fV})(a_{RQ}a_{QR})]$

Let $A = (a_{Vf}a_{fV})$, $B = (a_{RQ}a_{QR})$, and $C = (a_{QQ})$. Then, substituting the feedbacks into Equation 4.1 we get

$$(CA + CB)(-CA + d) + C^2 A^2 - 2CAd + d^2 < C^2(-AB)$$

Expanding and simplifying, we find the stability relations reduce to

$$B < A + \frac{d}{C}$$

Replacing the values of the links, we have the stability criteria in terms of the original graph

$$a_{RQ}a_{QR} < a_{Vf}a_{fV} + \frac{(a_{QV}a_{RQ}a_{VR} - a_{QV}a_{SQ}a_{VS})}{a_{QQ}}$$

As long as at-a-station river geometry and bedload transport have smaller interaction strengths than those between velocity and friction, we can expect stability. These conditions on the interactions are precisely those specified by Slingerland (1981). In addition, he cites the work of Andrews (Slingerland, 1981, p. 493), in which irrigation since the 1900s on Muddy Creek, Wyoming increased the sediment load to the East Fork River. First roughness and depth of the East Fork River increased with no noticeable change in slope, although this was anticipated. Hence $|a_{Vf}a_{fV}| > |a_{RQ}a_{QR}| > d$ and the result is consistent with the model requirements for the second stability condition. The fact that the first stability criterion is not met with $F_5 = 0$ means there is a conserved quantity. The difference between the two stability criteria not being met is evidently, for streams, the difference between a change in geometrical shape, as in the East Fork River versus the braiding of stream flow in glacial outwash areas.

A major controversy in ecology concerns the role of species interactions, and particularly competition, in determining the distribution and abundance of species. The individualistic hypothesis asserts that species occur where the physical conditions permit, essentially independent of each other. Therefore any correlation between the abundances or distributions of species would reflect correlations in their requirements rather than any interaction. Advocates of the role of competition, or more generally of community structure, claim that species do determine each other's distribution. In particular, presumed competitors might be expected to show negative correlations with each other. But observations often fail to confirm this expectation.

Also common sense is not always a reliable guide to what to expect. The diagrams of Figure 4.10 show competing species embedded in different contexts. The corresponding tables give the consequences of each.

In Figure 4.10a, we see a classical competition, a pair of competitors, both of which are self-damped. In the remaining cases, competition is mediated through the use of a common resource. In Figure 4.10b, the competitors have one resource in common; but one species also can use an additional resource. In Figure 4.10c, both species depend on the same single resource but one of them is preyed upon by a specialized predator. In Figure 4.10d, both competitors have predators, and in Figure 4.10e we have added a predator to the two-resource model of Figure 4.10b.

The first graph gives the expected results: any single parameter change results in A_1 and A_2 changing in opposite directions, therefore generating a negative correlation. In Figure 4.10b, three out of the four parameter changes generate negative correlations between the competitors, but inputs which affect the parameters of the common resource, R_2, change only A_2 and therefore do not produce correlations. Positive correlations may arise if, for example, R_1 and R_2 vary together but R_2 varies much more. Then the input

QUALITATIVE PREDICTIONS

(a)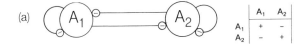

(b)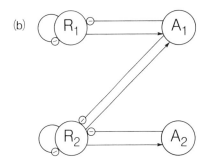

(c)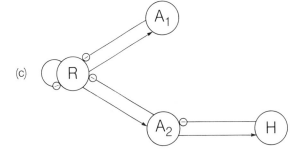

(d)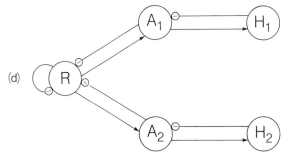

Fig. 4.10 *(continued overpage)*

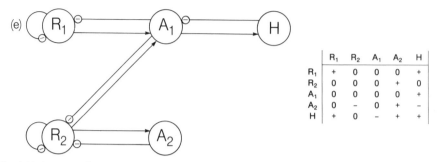

Fig. 4.10 (continued)

to R_2 has a positive effect on A_2 which can swamp the negative effect of the change entering at R_1, and both competitors increase.

Table 4.5 summarizes the following discussion.

Table 4.5 Turnover rates for Figure 4.10

Parameter change in—		Change in turnover rate			
		A_1	A_2		
Case (a)	A_1	0	0		
	A_2	0	0		
		R_1	R_2	A_1	A_2
Case (b)	R_1	+	0	0	0
	R_2	0	+	0	0
	A_1	+	0	(−)	0
	A_2	−	+	0	(−)

Look at case (a) of Table 4.5. There we see that an increase in A_1 is followed by a decrease in A_2 and a decrease in A_2 is met by an increase in A_1. There is no change in the turnover rate of either variable, just a change in equilibrium levels.

Now look at case (b). There we see that since we have increased the input rate to R_1 and the removal rate via A_1, the turnover rate but not the standing crop of R_1 has increased. The input to R_2 has not been changed, and neither has its equilibrium level. Therefore the increase in removal by A_1 exactly balances the reduced uptake by A_2, leaving the total removal rate unchanged. A_1 increases but its food supply and death rates have not been altered.

Similarly the turnover of A_2 is unchanged since its food supply and death rates are unchanged.

With input to R_2, there is no change in the turnover of R_1 since its rate of input and removal (A_1) are unchanged. R_2 turns over faster, since its input and its removal by A_2 have increased. A_1 and A_2 have no changes in food supply or death rates, so that although A_2 increases and A_1 remains at its previous level, the turnover rates of both are unchanged.

Parameter change at A_1 reduces its food supply R_1.* If the change acts on reproduction, compensating for the reduced food, there may be no change in turnover. But if we assume the parameter change is a reduction in death rate then turnover rate declines, with lower death rates balancing lower reproduction. The parenthesis in the table indicates this ambiguity. Since neither the resource nor death rate of A_2 changes with its lowered population size, turnover remains the same.

Parameter change at A_2 reduces A_1, so that the removal rate (and hence turnover) of R_1 decreases. R_2 is reduced. Since its input has not been changed, the removal rate must have increased (the changes in A_1 and A_2 are in opposite directions, but the net effect must result in increased removal of R_2). A_1 decreases. But since the death rate has not changed, the opposite effects of increased R_1 and decreased R_2 balance, and turnover is unchanged. Since A_2 increases despite reduced reproduction (reduced R_2), its turnover rate declines. If the parameter acted on reproduction there would be no change in turnover.

The table predicts that no single parameter change can generate any correlation in the turnover rates of A_1 and A_2. This might be observable in the field as the absence of correlation in average age or size. The turnover rates of A_1 and R_1 will be negatively correlated only if change enters at A_1. Change entering at A_2 produces opposite correlations of turnover rates with the two resources.

EXAMPLE 4.4

Find the turnover rates for the models of Figure 4.10c,d,e.

Solution

The solution is outlined in Table 4.6.

In this table we see that the $(+)$ for A_1 in response to input to A_1 holds if the parameter acts on the reproductive rate. Then both births and deaths

---

* An unchanged input but faster removal reduces R_1 and increases turnover. Since the level of R_2 is unchanged and its input is unchanged, its removal rate is also unchanged, the increase in A_1 balancing the decrease in A_2.

Table 4.6 Solution to Example 4.4

Parameter change in—		Change in turnover rate			
		R	A_1	A_2	H
Case (c)	R	+	0	0	0
	A_1	+	(−)	−	0
	A_2	0	0	(+)	0
	H	0	0	0	−

		R	A_1	A_2	H_1	H_2
Case (d)	R	0	+	+	0	0
	A_1	0	(+)	0	0	0
	A_2	0	0	0	0	0
	H_1	−	+	+	(+)	0
	H_2	−	+	+	0	(+)

		R_1	R_2	A_1	A_2	H
Case (e)	R_1	−	0	+	0	0
	R_2	0	+	0	−	0
	A_1	0	0	+	−	0
	A_2	0	0	−	−	0
	H	0	−	0	0	(−)

increase since H_1 increases. But if the parameter acts to reduce death due to causes other than H_1, there will be no change in turnover rate. □

Some General Observations

An examination of the tables corresponding to the graphs of Figure 4.10 notes the following:

There are many zeros in the graph. Although all the variables are connected directly or indirectly, they do not all respond to each kind of input. Whether a variable's equilibrium level responds or not depends on its position in the graph.

A subsystem of zero feedback linked to the rest of the system through a single variable only is referred to as a satellite of that variable. Thus in 4.10c, A_1 is a satellite of R and H is a satellite of A_2. R and A_2 are referred to as the principals of A_1 and A_2. A satellite buffers its principal from all parameter changes arising anywhere else in the system so that the column in the table for that principal will be all zeros, except for changes entering at the satellite. The satellite absorbs all the impact entering the system at its principal, so that the row in the table showing responses to parameter change at the principal will be all zeros, except for the satellite.

From the table corresponding to the graph of Figure 4.10c, we see that the turnover rate of a satellite is insensitive to parameter changes anywhere except at the satellite itself, since its only input from within the system is its principal, which changes only in response to the satellite.

A comparison of the graphs in 4.10 also show that the impact of a change in graph structure can be felt at places removed from that change. Thus the introduction of a second herbivore (going from 4.10c to 4.10d) changes the response of R, A_2, and H_2 to inputs to R. In the previous Figure 4.7e, the introduction of a larval predator T changes the effect of egg parameters on adult abundance. In the models of a green G and blue-green algal system (Figure 4.8a), the negative link from B to G causes the counterintuitive response of N to input at N, and in the diabetes model (Figure 4.8c) the A, E interaction determines the effects of inputs to G and I on each other.

The correlations among variables do not depend only on the relation of those variables themselves but also on the rest of the graph. Note that A_1 and A_2 are competitors for the resource R in 4.10c and d. In 4.10c they will show a negative correlation in response to parameter changes entering at H, but in 4.10d no single input will produce any correlation.

EXAMPLE 4.5

Show that in a stable system no variable can have more than one satellite.

Solution

A satellite variable is by definition without a self-effect loop and only connected with reciprocal interactions to its principal variable and none other in the system. In addition, it is only affected by the principal, although the satellite may itself have connections to other variables of the system. Thus, at the feedback level of the overall system, a principal variable with more than one satellite means that only one of those satellites can be on a loop with the principal variable. This leaves the remaining satellite variables isolated. They cannot contribute to the feedback of the overall system since they have no self-effect loops, so $F_n = 0$. Thus, no stable system can have a variable with more than one satellite. □

EXAMPLE 4.6

When does it become useful to introduce intervening variables, that is, variables that are implied by a link, but not explicitly stated. For example, show why an intervening physiological state of a consumer population might be introduced.

Solution

We already have looked at a two-variable system of a resource and consumer, as depicted in Figure 4.11. The rate of reproduction of X depends directly not

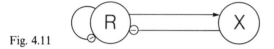

Fig. 4.11

on the resource in its environment but on its physiological state, which is fed by R. If we decide to include this state as an intervening variable between R and X, the graph will be the same as the one shown in Figure 4.12.

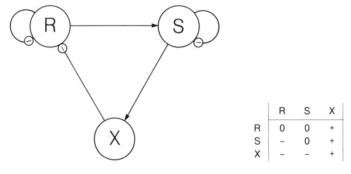

Fig. 4.12

	R	S	X
R	0	0	+
S	−	0	+
X	−	−	+

The utility of this decision depends on whether any new behavior would be observed in this expanded model. The effects of parameter change are shown in the table at the bottom of the figure.

As in the two-variable model without S, a change in a parameter which increases the growth rate of R increases the equilibrium value of X and leaves R unchanged. A change in parameter that increases the growth rate of X reduces the equilibrium value of R and increases X. Thus if a change in parameter enters at R, the impact will be seen only in the consumer population, while the physiological state and resource level are unchanged. Any parameter change in the physiological state S leaves S unchanged, but increases the population while reducing the resource level. Finally, a parameter change in the consumer population X increases its equilibrium population at the expense of a resource and results in a deterioration in the physiological state S.

These results hold for all equations consistent with the graph. Unlike the two-variable R, X system, it is possible for this system to show oscillatory instability if the links between variables are strong enough, compared to the self-damping of R and S. □

Resource management consists of intervening in a system by adding or removing some variables or changing parameters to meet some goal (for example, the harvesting of some species, or its protection). Sometimes our intervention takes the form of a parameter change such as increasing the death rate of fish. It also may change the structure of the graph.

Consider the simple three-variable system in the next figure (Figure 4.13).

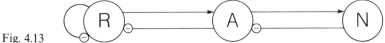

Fig. 4.13

Suppose first that we are interested in harvesting N. Several procedures may be adopted. One possibility is that there is a fixed harvest goal of so many tons per year. This rule gives N a positive self-effect. The original equation had the form

$$\frac{dN}{dt} = Nf(A)$$

At equilibrium $f(A) = 0$ so that

$$a_{NN} = \frac{\partial \frac{dN}{dt}}{\partial N}$$

which is $f(A)$, zero. But now

$$\frac{dN}{dt} = Nf(A) - H$$

for harvest H. Therefore the equilibrium level of N is

$$N^* = \frac{H}{f(A^*)}$$

and $f(A) > 0$.

The new graph structure and the corresponding table in contrast to the non-harvest model are shown in Figure 4.14.

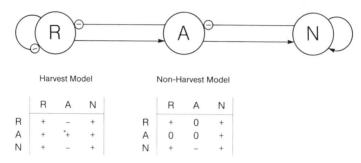

Fig. 4.14 *Assumes a stable system, therefore self-damping of R greater than self-enhancement of N

This management scheme changes the effects of the parameters of R and A. It also may destabilize the equilibrium since now

$$F_1 = -a_{RR} + a_{NN}$$
$$F_2 = +a_{RR}a_{NN} - a_{RA}a_{AR} - a_{AN}a_{NA} \quad \text{and}$$
$$F_3 = -a_{RR}a_{AN}a_{NA} + a_{NN}a_{RA}a_{AR}$$

any of which may be positive if the fixed harvest goal H is large enough. A second strategy would be to maintain a constant fishing rate, h. Then

$$\frac{dN}{dt} = N(f(A) - h)$$

The structure of the graph is unaltered and stable.

A third possibility is to harvest at a rate proportional to R since R provides the nutrients that eventually turn into N. We then have the graph shown in Figure 4.15.

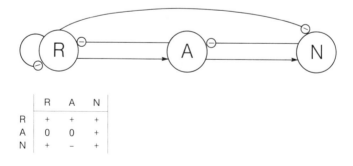

Fig. 4.15

If the new negative loop of length 3 is strong enough, this management scheme may destabilize the system. It also reverses the response of A to R from the previous model.

Lumping Variables

As we introduce more variables into the models in order to encompass complexity, computations become increasingly unwieldy and more of the predictions are ambiguous. Therefore it becomes important to have procedures for simplification, for the resolution of ambiguity, and for deriving inferences from only part of the graph.

A simplification procedure must leave intact the qualitative properties of interest. The two kinds of properties we examine are affected differently by the various procedures. Stability properties are more easily altered by a modification of the graph, while responses to parameter change can be preserved.

1. Grouping by rates of change. If we are interested in a particular set of variables, say invertebrates in a forest, the structure of the forest may be taken as given. That is, the population of trees can be treated as constant parameters on the time scale that invertebrate populations change. Similarly, those reactions which proceed much more rapidly than the variables of primary interest can be regarded as having already reached an equilibrium conditional on the variables of interest. For instance, bacteria which reproduce on a scale of hours and are consumed by invertebrates may reach an equilibrium determined by their own food supply and the predation on them. Their populations therefore will be expressions of invertebrate abundance, and these equilibrium levels can appear in the equations for invertebrates instead of their own populations. Likewise, if the physiological state of a consumer reaches equilibrium with the food supply much more rapidly than the population responds, then the physiological state need not appear as a distinct variable. Rather, population growth can be represented as a function of food supply (see the R, S, X system in Example 4.6). But then we can shift the focus, with the former fast variables becoming the variables of interest on an accelerated time scale and the former variables becoming constant parameters.

2. If a variable has only a one-way connection with the system of interest, it does not affect the loop structure which determines the dynamics. Then it is treated either as a parameter of other variables in the system (an input), or a product of the system.

3. A variable is an intermediate along a path. If it has no other links, it can be omitted from the model because its presence has no other qualitative influence on the response of the variables to parameter change. It may have to be included, however, if the stability properties of the system are being examined and its exclusion reverses the stability property of the graph. Thus what places variables in the same system is that they interact reciprocally on commensurate time scales.

4. A subsystem of K variables which is linked by one variable to the next may be replaced by a single variable. Then the self-damping of this new variable is the feedback F_K of the subsystem. Two particular cases are of special interest. A population can always be subdivided: we may distinguish animals in different patches, or in alternative physiological states that transform into each other, or engaged in different activities. In each of these situations, the total number of animals is unchanged as they move back and forth between categories—an increase in one implies a decrease in another. Since these categories are not self-reproducing—animals on the ground come from animals in the trees as well as the ground—each of these variables is self-damped. Nevertheless the movement between categories does not produce any net feedback for the subsystem as a whole. In Figure 4.16 we see hungry animals, H, feeding on some resource, R, and becoming satiated, S, while satiated animals metabolize and revert to being hungry.

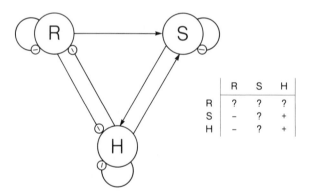

Fig. 4.16

The complement of R is the $[SH]$ system, in which F_2 is ambiguous due to positive feedback in the loop of length 2 and two negative loops of length 1. But in fact the feedback of the $[SH]$ system as produced by the transformation between hunger and satiation is zero. Therefore we can replace S and H by animals, A, and obtain the graph shown in Figure 4.17. In this graph we see that input to R has no effect on the equilibrium level of R but increases total A; inputs to A decrease R and increase A. We have gained certainty

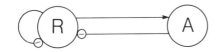

Fig. 4.17

about the R, A relations but have lost information about how the proportions of S and H change.

A similar situation arises when we divide a population into age or size-classes. Let X_i be the class i. Then

$$\frac{dX_0}{dt} = f_0 X_0 + f_1 X_1 + \cdots + f_K X_K$$

where the f's are fecundities, and

$$\frac{dX_i}{dt} = a_{i-1} X_{i-1} - b_i X_i \quad \text{for } i > 0$$

where a is a rate of transformation of the $(i-1)$st class into the ith class, and b is the rate of removal of class i by death or growth. Once again the separate variables X_i are self-damped with positive loops of lengths greater than one. The new variable X which lumps all classes is not self-damped. The general rule is, if for any set of X_i's the differential equations are linear and homogeneous in the X's (that is, are of the form $(dX_i/dt) = \sum a_{ij} X_j$, even if the a_{ij} are functions of variables other than the X's), then the subsystem of the X's has zero feedback.

5. If several variables in a set behave identically in their interactions with variables outside that set, then the problem breaks down into two parts: one which lumps all the variables in the set as a single new variable in its interactions with the rest of the system, and a separable problem analyzing the relations among variables within that set. If this absolute assumption "almost" holds, the problem "almost" breaks down in that way.

6. The original variables may be replaced by new variables which are more tractable and reveal the significance of changes more readily. In Figure 4.18 A_1 and A_2 have the same position in the graph.

A parameter change in N ("an input" to N) can have a positive effect on A_1 directly and a negative effect by way of A_2 and H, so that the response of A_1 to N is ambiguous. Similarly the effect on A_2 is ambiguous. Since input to N does not affect the level of H despite possible changes in both of its resources, the resource available to H, the linear combination $a_{H1} A_1 + a_{H2} A_2$ has not changed. Therefore we create the new variable

$$A_3 = a_{H1} A_1 + a_{H2} A_2$$

which is resource to H.

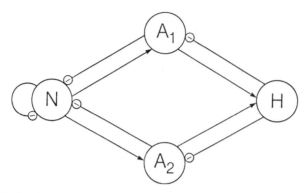

Fig. 4.18

Similarly, input to H does not affect N, although it can affect both consumers of N in ambiguous ways. Therefore we create the new variable

$$A_4 = a_{N1}A_1 + a_{N2}A_2$$

the combined algal consumer of N. A_4 increases when there is input to N since N remains unchanged.

At first glance the stability of the graph in Figure 4.18 is also ambiguous, since

$$F_4 = -[a_{N1}a_{1N}a_{2H}a_{H2} + a_{N2}a_{2N}a_{1H}a_{H1}]$$
$$+ [a_{N1}a_{2N}a_{1H}a_{H2} + a_{N2}a_{1N}a_{H1}a_{2H}]$$

If the two species of A_1 and A_2 are identical in their interactions, $F_4 = 0$. But if the species are different, it is plausible that when $a_{N1} > a_{N2}, a_{1N} > a_{2N}$, and so on. That is, if A_1 consumes more N than A_2, then N contributes more to A_1 than to A_2. In that case, let

$$a_{1N} = a_{XN} + \epsilon, \qquad a_{2N} = a_{XN} - \epsilon$$

and

$$a_{N1} = a_{NX} + \epsilon, \qquad a_{N2} = a_{NX} - \epsilon$$

and likewise for the H terms. Then

$$F_4 = -[(a_N + \epsilon)^2(a_H + \eta)^2 + (a_N - \epsilon)^2(a_H - \eta)^2]$$
$$= 2[(a + \epsilon)(a - \epsilon) + (a + \eta)(a - \eta)]$$
$$= 2[(a^2 - \epsilon^2)(a^2 - \eta^2)] - 2[(a^2 + \epsilon^2)(a^2 + \eta^2)] < 0$$

EXAMPLE 4.7

A subsystem of a larger model consists of two variables: nutrients N and a plant population P that uses the nutrient, as given in the model of Figure 4.19.

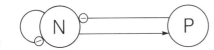

Fig. 4.19

We are interested in the relation of plants to the rest of the system. The inclusion of the nutrient as a separate variable complicates the analysis. Under the assumption that the nutrient has no other interaction to the rest of the system except via the plant population, show the consequences of lumping the nutrient variable N with the plant population by incorporating the equilibrium value of N into the variable P.

Solution

Suppose the nutrient is coming into the system through geological processes, labeled input, I, and removed from the system at some consumption rate times the abundance of plants and other factors in the environment

$$\frac{dN}{dt} = I - N(P + C_1)$$

The plants enter the system and change at a rate dependent on the number of plants times the input of nutrient minus some death rate due to predation

$$\frac{dP}{dt} = P(N - C_2) \qquad (E4.7.1)$$

In this system, N is self-damped but P is not.

The equilibrium value of N is

$$N^* = \frac{I}{P + C_1}$$

Substituting this value into Equation 4.7.1 we get

$$\frac{dP}{dt} = P\left(\frac{I}{P + C_1} - C_2\right)$$

and

$$\frac{\partial P'}{\partial P} = -\frac{I}{(P + C_1)^2}$$

There is a negative feedback in the system where it did not exist before. The self-damped variable at the bottom transfers its self-damping one link upward. This always will happen when lumping from the bottom of any system and moving upward.

There is an arbitrary choice of what variables to include based on the interest of the investigation. Consistency requires a price be paid. In the model above, the loop of length 2 is lost when the negative feedback on N via P becomes the self-damping on P. The phenomenon remains the same, however. □

Model Ambiguity

Most of the examples we have considered so far have been chosen to allow unambiguous decisions about the local stability of the systems shown. In real systems, however, the results are often less clear. Therefore we now consider several kinds of ambiguities that may arise. In order to do so, we sometimes will have to specify the mathematical forms of the interactions to a greater extent than previously. (Note: this section should be skipped by those unfamiliar with differential equations until reading Chapter 6.)

1. Loop combinations of opposite sign. Here we mean that there are loops which contribute to the same feedback level but are of opposite signs; hence the feedback sign is indeterminate. Sometimes further information can be obtained about the sign by making assumptions, but is short of quantification. Two cases are described.

Case 1. In a previous example we considered the model for a system with nitrogen, N, phosphorus, P, green and blue-green algae. The graph for the model is repeated here as Figure 4.20.

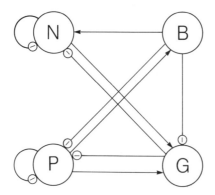

Fig. 4.20

Recall that the effect on G of a parameter change entering at B was ambiguous. The pathway BNG ($a_{NB}a_{GN}$) involving nitrogen fixation is positive, while the BPG ($a_{PB}a_{GP}$) and BG (a_{GB}) paths which correspond to competition for phosphorus and toxicity are negative. One conclusion we can reach is that if nitrogen fixation is strong enough, compared to the other processes, then a parameter change to B increases G. Another approach examines the role of variable abundances on the paths and complements. This requires some plausible assumptions about the equations for the variables. Suppose

$$\frac{dG}{dt} = G[a_{GN}N + a_{GP}P - a_{GB}B - m_G]$$

$$\frac{dB}{dt} = B[a_{BP}P - m_B]$$

$$\frac{dN}{dt} = I_N + a_{NB}B - N[a_{NG}G + C_N]$$

$$\frac{dP}{dt} = I_P - P[a_{PB}B + a_{PG}G + C_P]$$

where m_G and m_B are mortality rates and C_N and C_P are rates of removal from the system by physical and chemical processes that do not involve B and G. Then the pathway BNG contributes $a_{NB}a_{GN}G$ with complement $-(a_{PB}B + a_{PG}G + C_P)$. The pathway BPG is $-(a_{PB}P)(a_{GP}G)$ with complement $-(a_{NG}G + C_N)$. The path BG is $-a_{GB}G$ with complement $-(a_{PB}B + a_{PG}G + C_P)(a_{NG}G + C_N)$. Thus the combined effect, after division by the negative F_4, has the sign of

$$(a_{NB}a_{GN}G)(a_{PB}B + a_{PG}G + C_P) - (a_{PB}P)(a_{GP}G)(a_{NG}G + C_N)$$
$$- a_{GB}G(a_{PB}B + a_{PG}G + C_P)(a_{NG}G + C_N)$$

The first term is quadratic in the variables G and B while the negative terms are cubic in G and contain P as well. Therefore, if G or P is very large the overall effect is negative. Positive input to N increases G, eventually producing a positive impact of B on G.

Fig. 4.21

Case 2. The model in Figure 4.21 is that of an ordinary trophic hierarchy but, in addition, nutrient is recycled by the excretion or mortality of H. $F_1 < 0$, $F_2 < 0$ but F_3 is ambiguous:

$$F_3 = -\alpha_{NN}\alpha_{AH}\alpha_{HA} + \alpha_{AN}\alpha_{HA}\alpha_{NH}$$

Since $\alpha_{HA} > 0$ (A increases H), feedback at level three is determined by the sign of $-\alpha_{NN}\alpha_{AH} + \alpha_{AN}\alpha_{NH}$. To make the determination, we have to be more specific. One choice is

$$\frac{dN}{dt} = I - N[\lambda + \beta_{NA}q(N)A] + \beta_{NH}H$$

where I is a rate of inflow of N to the system, λ is an outflow rate, and $Nq(N)$ is some increasing function of N. Then

$$\alpha_{NN} \equiv \frac{\partial N'}{\partial N} = -\left[\lambda + \beta_{NA}A\frac{\partial Nq(N)}{\partial N}\right]$$

Similarly

$$\frac{dA}{dt} = A[\beta_{NA}Nq(N)] - \beta_{AH}s(A)H$$

where $s(A)$ is an increasing function of A. Then

$$\alpha_{AN} \equiv \frac{\partial A'}{\partial N} = \beta_{NA}A\frac{\partial Nq(N)}{\partial N}$$

Further

$$\alpha_{AH} \equiv \frac{\partial A'}{\partial H} = -\beta_{AH}s(A)$$

and finally

$$\alpha_{NH} \equiv \frac{\partial N'}{\partial H} = \beta_{NH}$$

Therefore F_3 has the sign of

$$-\left[\lambda + \beta_{NA}A\frac{\partial Nq(N)}{\partial N}\right][\beta_{AH}s(A)] + \beta_{NA}A\frac{\partial Nq(N)}{\partial N}\beta_{NH} \quad (4.2)$$

Because $s(A)$ is an increasing function of A, as A increases F_3 becomes more negative and the system thereby becomes more stable. For A near zero

$$F_3 \approx -\lambda\beta_{AH}\frac{\partial s(A)}{\partial A}\bigg|_{A=0} + \beta_{NA}\beta_{NH}$$

which can have either sign. Therefore all we can say is, for the appropriate constants (for instance, for a low flow-through rate, λ) the system may be unstable, but if A increases sufficiently it will be stable. Finally we return to the model of Figure 4.21 to determine how A may be increased. Inputs directly to N or A will not change A, whereas a positive input to H changes A in the direction of the sign of $\alpha_{AN}\alpha_{NH} - \alpha_{AH}\alpha_{NN}$, that is, it has the same sign as F_3 (Eq. 4.2). Therefore if we begin with a stable system, negative input to H can increase A and stabilize it further, while positive input to H may make the system unstable. Roughly, it looks as if high N and H and low A can result in instability.

2. The sign of a loop or path itself may be ambiguous. This type of ambiguity arises from opposing processes for which quantification cannot or has not been made. For instance, if a plant is self-inhibiting due to crowding and also is preyed on by an herbivore, as shown in Figure 4.22, it can be represented by Equation 4.3.

Fig. 4.22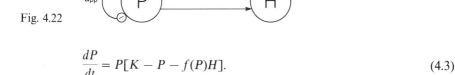

$$\frac{dP}{dt} = P[K - P - f(P)H]. \tag{4.3}$$

$Pf(P)$ is the functional response of predators to herbivores. If $Pf(P)$ is an upward convex function, such as

$$Pf(P) = \frac{vP}{K + P}$$

then

$$\frac{\partial P'}{\partial P} = \left[-1 + \frac{vHK}{(K + P)} \right] P$$

Thus for small P and large H this can be positive and the system may be unstable. In an initially stable system, positive input to P increases only H, so that an increase in the productivity of P will destabilize. Also, positive input to H (for example, reduction in its death rate) will reduce P and increase H. Either change therefore will destabilize. If the functional response is sigmoid (Figure 4.23), then below the inflection point the derivative is $\partial f/\partial P > 0$ and the system is stable; but above some threshold of P, $\partial f/\partial P < 0$,* and there will be instability.

* Figure 4.23 is a plot of $Pf(P)$. Since P is always increasing, to get the sigmoid shape (the leveling off) $f(P)$ has to decrease above the inflection point.

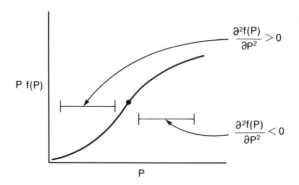

Fig. 4.23

EXAMPLE 4.8

In the model shown in Figure 4.24, G represents grass, M, mice, and C, coyotes.

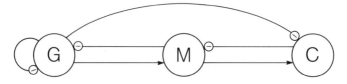

Fig. 4.24

	G	M	C
G	+	+	+
M	0	0	+
C	+	−	+

In addition to the feeding relations the grass affords some protection for the mice, thus having a direct negative link to coyotes. Let G be an ordinary self-damped population preyed on by mice without saturation

$$\frac{dG}{dt} = G[K - G - \beta_{GM}M]$$

Then, $\alpha_{GG} = -G$

$$\frac{dM}{dt} = M[\beta_{MG}G - \beta_{MC}Cs(G)]$$

where the β are constants and $s(G)$ reflects the protection grass offers to mice. Therefore $s(G)$ is a decreasing function of G and we have

$$\alpha_{MG} = \beta_{MG}M - \beta_{MC}CM\frac{\partial s}{\partial G} \quad \text{and}$$

$$\alpha_{MC} = -M\beta_{MC}s(G)$$

Finally

$$\frac{\partial C}{\partial t} = C[\beta_{CM} M s(G) - \theta]$$

where θ is the death rate of C. Therefore

$$\alpha_{CG} = \beta_{CM} CM \frac{\partial s}{\partial G} < 0$$

Explore the role of the three species and the strengths of the interactions to the stability of the model.

Solution

F_1, F_2, and F_3 are all negative. The possibility of instability depends on $F_1 F_2 + F_3 < 0$. This term is

$$\alpha_{GG} \alpha_{GM} \alpha_{MG} + \alpha_{CG} \alpha_{MC} \alpha_{GM}$$

Since α_{GM} is positive, the expression has the sign of

$$\alpha_{GG} \alpha_{MG} + \alpha_{CG} \alpha_{MC}$$

The stability function $F_1 F_2 + F_3$ has the sign of

$$-GM \left[\beta_{MG} - \beta_{MC} C \frac{\partial s}{\partial G} + \beta_{CM} CM \frac{\partial s}{\partial G} (-M \beta_{MC} s(G)) \right]$$

Since $\partial s / \partial G$ is negative, this leaves

$$GM \beta_{MC} C \left| \frac{\partial s}{\partial G} \right| - GM \beta_{MB} + M^2 C s(G) \left| \frac{\partial s}{\partial G} \right| \beta_{CM} \beta_{MC}$$

which has the sign of

$$C \left| \frac{\partial s}{\partial G} \right| \{ \beta_{MC} G + M s(G) \beta_{CM} \beta_{MC} \} - G \beta_{MG}$$

Since the coefficient of coyotes, C, is positive, increases in C would have a stabilizing effect. Thus a positive parameter input to the mice promotes stability. But it is not possible to change M and C, without changing C and G as well. We can guess that positive parameter input to grass, G, is more likely to stabilize than input to coyotes, which changes M and C in opposite directions. The role of grass, G, is less obvious. The saturation curve, $s(G)$, increases less rapidly than G. Therefore for large G, the expression is (roughly)

$$G\left\{C\left|\frac{\partial s}{\partial G}\right|\beta_{MC} - \beta_{MG}\right\}$$

As s increases, the first term gets smaller and the whole term becomes negative. Therefore it seems as if increase in grass destabilizes while in mice and coyotes it stabilizes. But we can say further that because $|\partial s/\partial G|$ gets smaller as G increases, input to G, which increases all three variables, most likely stabilizes by increasing M and C more than it destabilizes by way of G. Further precision would require making more specific assumptions about the functions. □

When we compare the results of the different models in the two cases of the first section of this chapter and the preceding example, we see there is no general rule which states that increased productivity at the plant level necessarily stabilizes or destabilizes. The outcome would depend on the structure of the community.

This analysis suggests some curious evolutionary consequences. In cases where there is a parameter change that increases the rate of growth of the plant population (improved efficiency in photosynthesis, tolerance to physical conditions, or better seed set), evolution within the plant community can destabilize the system and lead to the local extinction of the plant. Survival therefore depends on immigration from other communities where selection has not proceeded that fast (Levins, 1975).

Changes of variable or plausible assumptions beyond the signed-digraph can remove some ambiguities; others will remain. As the number of variables in the graph increases, the proportion of predictions which are ambiguous increases. In a graph with twenty variables and four hundred entries, if even 5 percent of the predictions (twenty predictions) are unambiguous, this still would allow for strong testing of the model.

Experimental Design and Loop Models

A model represents our knowledge of one aspect of the world. The goal of qualitative modeling is to develop models to increase that knowledge. Verification of qualitative models is not an attempt to identify or perfect any single model as the perfect model of nature. Therefore we are looking for experimental design and field observations that amplify where and why a set of models are the same and different in their predictions.

The strength of the qualitative, loop models is their independence of quantification of absolute link strengths. This means that any prediction is relatively weak, being a plus, zero, or minus. Moreover, some predictions will be ambiguous because *relative strengths* of links will determine the sign of a prediction. There will be occasions when quantification is needed. A comparison of management plans to increase fish yield will require not only understanding that the plans produce the desired increase with roughly corresponding effects to the rest of the community, but also *how much* each plan is expected to achieve.

There are three ways to compare qualitative models to the natural world. The first way is to compare the pattern of correlation among the variables with that of the model predictions. The second way is to compare correlations among variables from places of known parameter change. The third way is to conduct a set of experiments in which the parameters of specified variables are changed and the outcome compared with the model predictions. These three ways both compare models with nature and expand our knowledge of the world. They will be described in greater detail in the remainder of this chapter.

We will describe a series of seven models for the high tide pools of New England. Although these models are neither inclusive nor necessarily definitive (see Puccia and Pederson, 1983), they illustrate the potential use of loop models in guiding data acquisition. For the sake of exposition some unrealistic assumptions about pool ecology have been included.

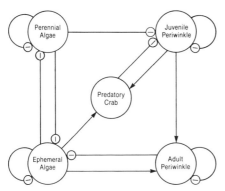

Model 1: Descriptive Model

	P	E	J	A	C
Perennial (P)	(+)	−	+	(+)	(−)
Ephemeral (E)	−	+	−	(−)	+
Juvenile Periwinkle (J)	0	0	0	0	(−)
Adult Periwinkle (A)	+	−	+	(+)	−
Crab (C)	−	+	(−)	(−)	(+)

Fig. 4.25

92 QUALITATIVE PREDICTIONS

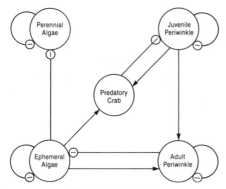

Model 2: No Algal Competition

	P	E	J	A	C
Perennial (P)	(+)	0	0	0	0
Ephemeral (E)	−	+	−	(−)	+
Juvenile Periwinkle (J)	0	0	0	0	(−)
Adult Periwinkle (A)	+	−	+	+	−
Crab (C)	−	+	−	−	+

Fig. 4.26

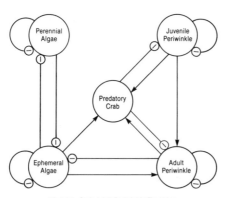

Model 3: Crab Adult Periwinkle Predation

	P	E	J	A	C
Perennial (P)	(+)	−	(+)	(+)	−
Ephemeral (E)	−	+	−	(−)	+
Juvenile Periwinkle (J)	−	+	(−)	(−)	(+)
Adult Periwinkle (A)	+	−	(+)	(+)	(−)
Crab (C)	−	+	(−)	(−)	(+)

Fig. 4.27

The first three models (Figures 4.25–4.27) show the important macroorganism variables for the tide pools: the adult (A) and juvenile (J) periwinkle snail, *Littorina littorea*; perennial algae (P), primarily *Chondrus crispus*; and ephemeral algae (E), including *Dumontia, Ulva,* and *Ceramium*; and the juvenile predatory crab (C), *Carcinus maenas*. In each of these models the interactions between the species is either from field observations (Pederson and Puccia, 1982) or as reported in the literature (Lubchenco, 1978; Sze, 1982).

For each model a table of predictions is given. An item in parenthesis means the result is ambiguous and requires additional assumptions about relative strengths of links or the stability characteristics of a complementary subsystem. These details would only clutter the present discussion; therefore, the tables should be taken as given. Anyone inclined to derive the tables of predictions, however, can resolve most of the ambiguities by assuming in models 1 and 3 that the perennial-ephemeral subsystem is stable (hence negative feedback) while in all other models assume it is unstable (positive feedback).

Models 4, 5, and 6 (Figures 4.28–4.30) delete the crab as a variable, replacing it with microflora-encrusting algae. Consequently, the last column and row do not correspond to the crab variable of the previous three models, invalidating any comparison for these columns and rows. The final model 7 (Figure 4.31) includes both microflora and crab as variables.

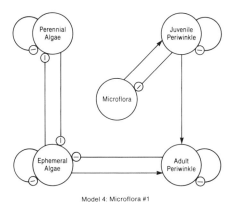

Model 4: Microflora #1

	P	E	J	A	M
Perennial (P)	+	−	0	−	0
Ephemeral (E)	−	+	0	+	0
Juvenile Periwinkle (J)	0	0	0	0	(−)
Adult Periwinkle (A)	+	−	0	(−)	0
Microflora (M)	+	−	+	(−)	(+)

Fig. 4.28

94   QUALITATIVE PREDICTIONS

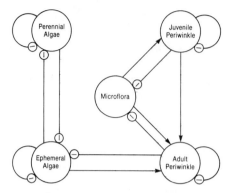

Model 5: Microflora #2 — Common Resource

	P	E	J	A	M
Perennial (P)	+	−	+	−	+
Ephemeral (E)	−	+	−	+	−
Juvenile Periwinkle (J)	−	+	(−)	(+)	(−)
Adult Periwinkle (A)	+	−	(+)	(−)	(+)
Microflora (M)	+	−	+	(−)	(+)

Fig. 4.29

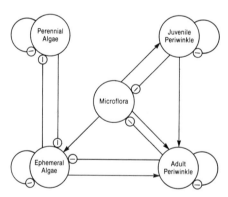

Model 6: Microflora #3 — Enhance Ephemerals

	P	E	J	A	M
Perennial (P)	+	−	+	−	+
Ephemeral (E)	−	+	−	+	−
Juvenile Periwinkle (J)	(−)	(+)	(+)	(+)	(+)
Adult Periwinkle (A)	+	−	(+)	(−)	(+)
Microflora (M)	(+)	(−)	(+)	(−)	(+)

Fig. 4.30

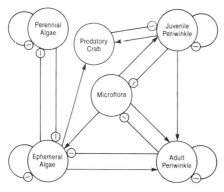

Model 7: Microflora and Crab

	P	E	J	A	C	M
Perennial (P)	(+)	−	+	−	(−)	(−)
Ephemeral (E)	−	+	−	+	(+)	+
Juvenile Periwinkle (J)	0	0	0	0	(−)	0
Adult Periwinkle (A)	+	−	+	−	(−)	(−)
Crab (C)	(−)	(+)	(+)	(−)	(+)	(+)
Microflora (M)	(+)	(−)	(+)	(−)	(+)	(−)

Fig. 4.31

For every model any sign in the first four columns or rows can be compared directly with the prediction for model 1; likewise, signs under a column or row labeled *C* indicate how the model under study compares with model 1; but the signs in any column labeled *M* signify differences between that model and model 4. In some of the entries in the tables there are signs in parentheses. This means that additional assumptions have been made in order to determine the sign. For example, in model 1 we assume that consumption of ephemerals by adult snails is a stronger interaction than protection given by ephemerals to crabs or the strength of maturation of juvenile snails into adults. For the present, the exact nature of these assumptions is unimportant. They enter into the discussion in a serious way when the nature of tide pool dynamics needs to be understood, which is beyond the scope of this presentation.

Correlation Patterns

The first way to differentiate among the models is to examine field observations of the tide pools in different geographic areas. Over a sufficiently wide and diverse geographic range there will be natural variation causing parameter change at one or more of the nodes represented in each signed-digraph model. There may be an increase in salinity, temperature, pollution, or enrichment in different areas, which may increase the growth rate of

Chondrus (*P*) or decrease the mortality rate of ephemeral algae (*E*). From the graph analysis point of view, this is the same as a parameter change (either + or −) at each row in the table of predictions. Therefore, looking down any two columns in the table yields pairwise patterns of covariation. For example, in model 1 the perennial (*P*) and ephemeral (*E*) algae either always change in opposite direction or have no response. Thus they always show a negative correlation; a comparison of the columns for juvenile snails (*J*) and crabs (*C*) shows they covary in the opposite direction—when one increases the other decreases; therefore we see they are negatively correlated. An examination of the ephemeral (*E*) and crab (*C*) columns produces a positive correlation—an increase or decrease in *E* is always paralleled with a rise or fall in *C*, respectively.

The pattern of covariation for all possible pairwise associations can be determined for each tide pool model table of prediction. Some cases will be indeterminate; for example the juvenile (*J*) and adult (*A*) periwinkle snails of model 1 in general will rise and fall together, except when a change in parameters affects the ephemeral algae, in which case they will be negatively correlated; a change in parameter entering the juvenile has no effect on any variable other than crab, the satellite variable. Some pairwise patterns show zero correlation, but for simplicity we ignore zero in constructing our table of correlations. All possible correlation patterns are tabulated in Table 4.7a and summarized in Table 4.7b.

All the models predict a negative correlation of perennial and ephemeral algae. This is a robust conclusion because it is independent of the different assumptions for each of the models. Should field data indicate a positive correlation exists, then all the models become suspect. In fact, the negative correlation arises from the assumption of competition between the two algae groups and the isolation of the perennial algae from the rest of the community, except through its connection to the ephemeral algae. Another explanation for a positive correlation is if both groups of algae were undergoing changes in parameters that increased (or decreased) their growth rate simultaneously and strongly so as to temporarily overwhelm the network response. The models can be used to search for interactions that were not noticed before.

The summary of correlation patterns also can help plan a research study. Since it is impossible to observe or measure everything, the best choice is to gain maximum information, to obtain data that will help distinguish among the models. In this case the strategy is to collect data that divide the model predictions into half; therefore, half the models will be in disagreement with the data. Table 4.7b does not divide evenly; for the juvenile to adult (*J*/*A*) correlation pattern, as one example, we find three out of seven models predict a positive relation, and three out of seven a negative, with one model being

Table 4.7(a) Predicted correlation pattern for all models with variation entering across all variables

Model	P/E	P/J	P/A	P/C	E/J	E/A	E/C	J/A	J/C	A/C	P/M	E/M	J/M	A/M
1	−	+	+	−	−	−	+	+	−	−				
2	−	+	+	−	−	−	+	+	−	−				
3	−	+	+	−	−	−	+	+	−	−				
4	−	+	−		−	+		−			+	−	+	−
5	−	+	−		−	+					+	−	+	−
6	−	?	−	?	?	?	?	?			?	?	+	?
7	−	?	?	?	?	?	?	−	?	?	−	+	?	?

Table 4.7(b) Proportion of models predicting the same correlation pattern among the variables across observations

Sign of correlation	P:E	P:J	P:A	P:C	E:J	E:A	E:C	J:A	J:C	A:C	P:M	E:M	J:M	A:M
+	0	5/7	3/7	0	0	3/7	3/4	3/7	0	0	2/4	1/4	3/4	0
0	0	0	0	0	0	0	0	0	0	0	0	0	0	0
−	7/7	0	3/7	3/4	5/7	3/7	1/4	3/7	3/4	3/4	1/4	2/4	0	2/4
?	0	2/7	1/7	1/4	2/7	1/7	1/4	1/7	1/4	1/4	1/4	1/4	1/4	2/4

98   QUALITATIVE PREDICTIONS

ambiguous. Then one data collection strategy is to conduct a survey over a wide geographic coastal area that determines the population size of juvenile and adult periwinkle snails. Nearly all the models (five out of seven) predict a positive correlation of perennial algae and juvenile periwinkle snails (P/J). The two remaining models are ambiguous: these are both microflora models (models 6 and 7) with the question mark occurring because the perennial (P) and juvenile (J) columns associated with the crab (C) or adult (A) snail row show opposite correlation patterns than the other rows. Reading Table 4.7a by way of the ambiguities (the ? row) can help identify at what entry point "natural" variation enters. For data showing P/J with a negative correlation, in model 6 this means that parameter change enters at juvenile periwinkle snails in order for this model to correspond to the data; in model 7, the parameter change would be through the crab. Notice that if this field observation is correct, it would cast doubt on the validity of all the remaining models.

EXAMPLE 4.9

In many systems it is arguable that the likelihood of parameter changes at all nodes across some ensemble of samples are equal, probable, or possible. Moreover, it may be known that parameter changes brought on by natural variation occur in one or a limited number of identifiable variables. For the tide pool models, show the proportion of predicted correlation patterns among the models when it is known that the natural variation in parameter change happens only in the population growth rate of ephemeral algae (E).

Solution
The correlation pattern predicted for each of the seven models when parameter change enters only through ephemerals is determined by a pairwise comparison along the ephemeral (E) row. In model 1, for example, the ephemeral row is

	P	E	J	A	C
→E	−	+	−	(−)	+

The associated correlation pattern is

$P/E = -$	$P/J = +$	$P/A = +$	$P/C = -$	$E/J = -$
$E/A = -$	$E/C = +$	$J/A = +$	$J/C = -$	$A/C = -$

The same procedure can be used for each model, summarized in Table 4.8.

Table 4.8 Predicted correlation pattern for all models with parameter changes at ephemeral algae only

	P/E	P/J	P/A	P/C	E/J	E/A	E/C	J/A	J/C	A/C	P/M	E/M	J/M	A/M
Model 1 → E	−	+	+	−	−	−	+	+	−	−				
Model 2 → E	−	+	+	−	−	−	+	+	−	−				
Model 3 → E	−	+	+	−	−	−	+	+	−	−				
Model 4 → E	−	0	−		0	+		0			0	0	0	0
Model 5 → E	−	+	−		−	+		−			+	−	+	−
Model 6 → E	−	+	−		−	+		−			+	−	+	−
Model 7 → E	−	+	−	−	−	+	+	−	−	+	−	+	−	+

Table 4.9 Proportion of models predicting the same correlation pattern for parameter change at ephemerals only

Sign of correlation	P:E	P:J	P:A	P:C	E:J	E:A	E:C	J:A	J:C	A:C	P:M	E:M	J:M	A:M
+	0	6/7	3/7	0	0	4/7	4/4	3/7	0	1/4	2/4	1/4	2/4	1/4
−	7/7	0	4/7	4/4	6/7	3/7	0	3/7	4/4	3/4	1/4	2/4	1/4	2/4
0	0	1/7	0	0	1/7	0	0	1/7	0	0	1/4	1/4	1/4	1/4

From Table 4.8 the proportion of models with the same predicted sign can be calculated, as given in Table 4.9.

Again all the models predict a negative correlation of perennial algae (P) with ephemeral algae (E). The majority of models (six out of seven) predict a positive correlation of perennial algae with juvenile periwinkle snails (J) and negative correlation of ephemeral algae with juvenile snails. The single exception to the six models is the zero correlation of model 4. This is because the microflora (M) is a satellite of the juvenile periwinkle with no link to any other variable. Therefore, any change in parameters of ephemerals or any other variable other than microflora cannot affect the abundance of juveniles since all pathways leave the microflora as an isolated zero feedback in the complementary subsystem. □

Example 4.9 illustrates the use of a correlation pattern when it is known where parameter changes will occur. Sometimes the only way to know which variable has changing parameters is to conduct a manipulation experiment —cause the change in parameter in a specific variable, such as increasing the mortality rate or increasing the input rate.

A parameter change in any variable produces results that correspond to a specific row in the table of predictions for each model. Removing perennial algae from tide pools is equivalent to increasing the mortality rate of perennial algae; it is a parameter change that reduces the growth rate of perennials. For model 1 this produces the predictions (from the table)

	P	E	J	A	C
⊖P	−	+	−	−	+

(Reminder: The signs are reversed because the original table of model 1 was calculated for a change in parameter that increases the growth rate of the variables.) The correlation pattern for this row is

$P/E = -$ $P/J = +$ $P/A = +$ $P/C = -$ $E/J = -$
$E/A = -$ $E/C = +$ $J/A = +$ $J/C = -$ $A/C = -$

The same correlation pattern is produced for a change in parameter that increases the growth rate of perennials ($\rightarrow P$) because all the signs are reversed.

For the first six tide pool models there are five possible choices for experimental manipulation: perennial algae (P), ephemeral algae (E), juvenile periwinkle snails (J), adult snails (A), crabs (C) or microflora (M). Model 7 includes all six variables as candidates for experimental intervention. Each row of every model depicts a correlation pattern. Since we can intervene with either a positive or negative change to parameters we call this variation entering at a variable. Hence, for each experimental variation introduced in the five variables of model 1 there will be five sets of correlation patterns, not necessarily different, and likewise for all other models except model 7, which will have six sets. The correlation sign pattern for all the models is given in matrix form in Table 4.10.

For each model we can find out how many differences in predicted correlation patterns can occur when experimental manipulation is performed in any two variables. We hope to maximize the number of different predictions between any two experiments. For example, an experimental manipulation of perennial algae (P) and juvenile snails (E') will, according to model 6, produce the same correlation pattern except in these six:

$$P/J = \begin{cases} +P \\ -J \end{cases} \quad P/M = \begin{cases} +P \\ -J \end{cases} \quad E/J = \begin{cases} -P \\ +J \end{cases}$$

$$E/M = \begin{cases} -P \\ +J \end{cases} \quad J/A = \begin{cases} -P \\ +J \end{cases} \quad A/M = \begin{cases} -P \\ +J \end{cases}$$

The fact that there is widespread self-consistency within the rows of each model, excluding models 6 and 7, means not much is gained in the attempt to validate any one of them by experimental manipulation of more than one variable. The greater the differences between rows in the table of predictions, the more information obtained from alternative experimental manipulation. A comparison of the number of different correlation patterns predicted within each model for experimental manipulations choosing any pair of variables is summarized in Table 4.11. The large number of zeros substantiates the previous statement about the inefficacy of alternative experiments.

A set of alternative models such as those in Figures 4.24 to 4.30 and the corresponding tables of predictions can be used in several ways. The most straightforward use is to compare observations to the models in order to choose among them, as just described. Sometimes the observations may even be incompatible with all models and therefore may indicate the existence of unidentified variables which were not included. For instance, consider the table and signed digraph of the four variables N, A_1, A_2, and H, shown in Figure 4.32. (The table is from observations and not derived from the model.)

The A_2 row and column is a giveaway. It is not possible in a stable system to have a complete row or column of zeros. Zeros arise from the presence of satellite variables. But in this case, input to the satellite changes the equilibrium value of its principal, and input to the principal alters the satellite. Furthermore, a single principal cannot have two distinct satellites

Table 4.10 Correlation between variables due to variation

		P					E					J					A					C/M			
		P	E	J	A	C	P	E	J	A	C	P	E	J	A	C	P	E	J	A	C	P	E	J	A	C
Model 1	P	1	−	+	−	+	1	−	+	−	+	1	0	0	0	0	1	−	+	−	+	1	−	+	−	+
	E		1	−	−	+		1	−	−	+	0	1	0	0	0		1	−	−	+		1	−	−	+
	J			1	+	−			1	+	−	0	0	1	0	0			1	+	−			1	+	−
	A				1	−				1	−	0	0	0	1	0				1	−				1	−
	C					1					1	0	0	0	0	1					1					1
Model 2	P	1	0	0	0	0	1	−	+	−	+	1	0	0	0	0	1	−	+	−	+	1	−	+	−	+
	E	0	1	0	0	0		1	−	−	+	0	1	0	0	0		1	−	−	+		1	−	−	+
	J	0	0	1	0	0			1	+	−	0	0	1	0	0			1	+	−			1	+	−
	A	0	0	0	1	0				1	−	0	0	0	1	0				1	−				1	−
	C	0	0	0	0	1					1	0	0	0	0	1					1					1
Model 3	P	1	−	+	−	+	1	−	+	−	+	1	+	−	+	−	1	−	+	−	+	1	−	+	−	+
	E		1	−	−	+		1	−	−	+		1	+	−	−		1	−	−	+		1	−	−	+
	J			1	+	−			1	+	−			1	−	+			1	+	−			1	+	−
	A				1	−				1	−				1	−				1	−				1	−
	C					1					1					1					1					1

	P					E					J					A					C/M				
	P	E	J	A	M	P	E	J	A	M	P	E	J	A	M	P	E	J	A	M	P	E	J	A	M
Model 4																									
P	1	–	0	–	0	1	–	0	–	0	1	0	0	0	0	1	–	0	–	0	1	–	+	–	+
E		1	0	+	0		1	0	+	0		1	0	0	0		1	0	+	0		1	–	+	–
J			1	0	0			1	0	0			1	0	0			1	0	0			1	–	+
A				1	0				1	0				1	0				1	0				1	–
M					1					1					1					1					1
Model 5																									
P	1	–	+	–	+	1	–	+	–	+	1	–	+	–	+	1	–	+	–	+	1	–	+	–	+
E		1	–	+	–		1	–	+	–		1	–	+	–		1	–	+	–		1	–	+	–
J			1	–	+			1	–	+			1	–	+			1	–	+			1	–	+
A				1	–				1	–				1	–				1	–				1	–
M					1					1					1					1					1
Model 6																									
P	1	–	+	–	+	1	–	+	–	+	1	–	+	+	+	1	–	+	–	+	1	–	+	–	+
E		1	–	+	–		1	–	+	–		1	+	+	+		1	–	+	–		1	–	+	–
J			1	–	+			1	–	+			1	–	+			1	–	+			1	–	+
A				1	–				1	–				1	–				1	–				1	–
M					1					1					1					1					1

(continued overpage)

Table 4.10 (continued)

		P						E						J				
		P	E	J	A	C	M	P	E	J	A	C	M	P	E	J	A	C	M
Model 7	P	1	−	+	−	−	−	1	−	+	−	−	−	1	0	0	0	0	0
	E		1	−	+	−	+		1	−	+	−	+		1	0	0	0	0
	J			1	−	+	−			1	−	+	−			1	0	1	0
	A				1	+	+				1	+	+				1	0	0
	C					1	+					1	+					1	0
	M						1						1						1

		A						C						M				
		P	E	J	A	C	M	P	E	J	A	C	M	P	E	J	A	C	M
Model 7	P	1	−	+	−	−	−	1	−	−	+	−	−	1	−	+	−	+	−
	E		1	−	+	−	+		1	+	−	+	+		1	−	+	−	+
	J			1	−	+	−			1	−	+	−			1	−	+	−
	A				1	+	+				1	+	+				1	+	−
	C					1	+					1	+					1	−
	M						1						1						1

QUALITATIVE PREDICTIONS   105

Table 4.11 Number of different predicted correlations from experimental intervention in pairs of variables

	Model 1					Model 2					Model 3			
	E	J	A	C		E	J	A	C		E	J	A	C
P	0	10	0	0	P	10	0	10	10	P	0	0	0	0
E		10	0	0	E		10	0	0	E		0	0	0
J			10	10	J			10	10	J			0	0
A				0	A				0	A				0

	Model 4					Model 5					Model 6			
	E	J	A	M		E	J	A	M		E	J	A	M
P	0	3	0	7	P	0	0	0	0	P	0	6	0	0
E		3	0	7	E		0	0	0	E		6	0	0
J			3	10	J			0	0	J			6	6
A				7	A				0	A				0

	Model 7				
	E	J	A	C	M
P	0	15	0	8	5
E		15	0	8	5
J			15	15	15
A				8	5
C					9

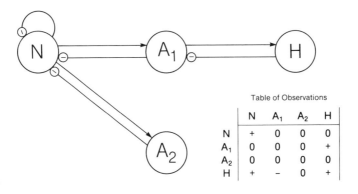

Table of Observations

	N	A_1	A_2	H
N	+	0	0	0
A_1	0	0	0	+
A_2	0	0	0	0
H	+	−	0	+

Fig. 4.32

since they would result in $F_n = 0$ (see Example 4.5). The satellites are included in F_n only by way of the principal, so that in any F_n one of them is left disconnected. Therefore the zero row and column suggest that A_2 has a satellite, H_2. When H_2 is included, A_2 is no longer a satellite of N and the augmented table is consistent. Conversely, if inputs to A_2 affected N and H but inputs from N to other variables were not zero, then A_2 neither has a satellite nor is a satellite of N. This suggests either that A_2 is self-damped or is linked to a self-damped variable or system, such as an additional nutrient N_2. From the perspective of the N, A_1, A_2, H system we cannot distinguish among these alternatives. In fact, if we know that A_2 links this subsystem to the larger community (which we are not working with), we can condense all this influence into a self-damping for A_2 and study the subsystem alone.

A second use of the set of graphs is in experimental design. Experiments consist of interventions in the system, the observation of outcome (in these cases the changes in equilibrium level or turnover), and the comparison of these results with the postulated graphs. If we could apply inputs to each variable in turn and wait for the system to equilibrate, we would have a complete table. From the complete table we could work backward to the graph, as will be explained in Chapter 5. But this may not always be possible. Resources or time may not permit the complete experiment. Or we may lack a suitable technique for targeting particular variables.* Then an examination of the graph may indicate either that accessible procedures giving us one or a few rows of the table may be sufficient to decide among alternatives, or that that will be insufficient and modification of the graph is necessary.†

The graph may be modified in the following ways:

1. A variable may be removed from the dynamics of the system and converted into a constant parameter. This does not mean that it has been removed physically. If we monitor the system closely, and either remove or add to a particular variable rapidly enough, it can be held effectively constant. Suppose that in an aquarium tank we want to remove snails as variables. But we may still want the snails there to scrape the algae. Then our procedure would be to have an auxiliary colony of snails available. When snails in the experimental tank die they are quickly replaced, while newborn snails are removed; the snail population remains constant. The snails still eat what they ate before, leave the same wastes, and behave in the same way. But they are no longer a variable, with numbers changing in response to other variables and, in turn, transmitting the received changes to other members of the network.

---

* Or, for measuring the outcome; there may be too many ambiguities in the table to be able to differentiate clearly among models.
† Also for further verification.

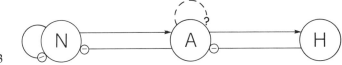

Fig. 4.33

Consider the graph shown in Figure 4.33. We do not know if A has self-damping or self-excitation. One possibility would be to remove H. The remaining graph is shown in Figure 4.34. The change in N when there is a

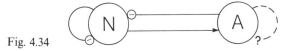

Fig. 4.34

parameter change in the growth rate of N would be $+$, 0, or $-$, depending on whether the loop a_{AA} is negative, zero, or positive. The system then may become unstable, since feedback at Level 1 or 2 could be positive. It is important, however, to consider how to remove H. If the original equation for A were

$$\frac{dA}{dt} = A\left(a_{NN}N - \frac{pH}{K+A} - \theta_A\right)$$

where θ_A is the death rate of A due to causes other than H. The self-loop would be

$$\frac{\partial}{\partial A}\left(\frac{dA}{dt}\right) = \frac{+pAH}{(K+A)^2}$$

a positive feedback. If H were removed physically from the system ($H = 0$), the positive feedback also would disappear and the experiment would destroy its object. If H is removed as a variable by holding it constant, then the positive feedback is retained.

The destabilization of a community is one way of revealing its structure. If two competitors share a common resource they may coexist because a predator utilizes one or both of them, as illustrated in Figure 4.18, reproduced here as Figure 4.35.

The removal of H would leave a system with $F_3 = 0$ so that one of the competitors would disappear. H has been labeled a *keystone predator* (Paine, 1966), and the exclusion of H (a starfish) confirmed the expectation in one case.

Another way to remove a variable is by saturation. Nitrogen is a necessary nutrient for phytoplankton. But if a great abundance of nitrogen saturates the capacity of algae to utilize the phosphorus, then changes in the nitrogen

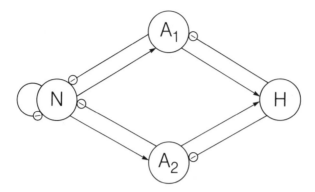

Fig. 4.35

level no longer have any effect. Nitrogen has been removed effectively from the network although it is still there physically.

In the system shown in Figure 4.36, the saturation of N causes it to leave the system, producing the graph of Figure 4.37, which is unstable ($F_3 = 0$). Either B or G will be displaced, depending on which one can grow on lower phosphorus levels.

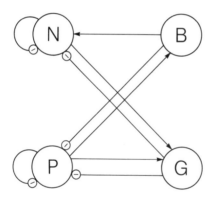

Fig. 4.36

	N	P	B	G
N	0	0	−	+
P	0	0	+	+
B	+	−	+	?
G	−	−	−	+

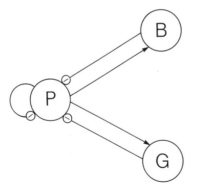

Fig. 4.37

2. New variables may be added to the system. Sometimes this can be done by the introduction of a species into an experimental community or of a chemical into a physiological preparation. But we also could introduce a pseudovariable, a computed variable which changes according to some rule we prescribe. For example, consider the system shown in Figure 4.38. In

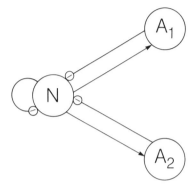

Fig. 4.38

order to stabilize the system we defined a pseudopredator P as follows:

$$\frac{dP}{dt} = (qA_1 - c)P$$

and intervene to remove A_1 at rate qPA_1. We thereby have changed the graph to appear as Figure 4.39.

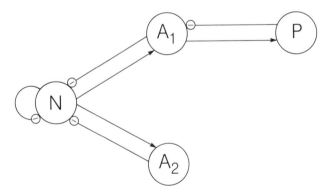

Fig. 4.39

Sometimes attempts at regulation can be understood as the addition of variables. Suppose that in some lake a fish population, F, is fished by industry at a rate, I, that depends on the price of fish and its abundance. This may give us the graph depicted in Figure 4.40, where the self-damping of F contains all the relations of the fish to its own food supply and predators. Increased

Fig. 4.40

fishing effort clearly reduces the fish population, while increased reproduction or survival of fish will encourage more fishing.

Now suppose a regulatory agency is established to protect the fish. The amount of government intervention is reduced when fish are abundant, but increases when fish decline. Moreover, the agency responds to its own estimate of fish population which is some moving average, say

$$\text{Estimate } (F) = aF_t + bF_{t-1} + cF_{t-2} + \cdots$$

for fish caught on different dates (that is, $(t-1)$, $(t-2)$, $\cdots$).

The agency may respond to fish decline by inhibiting fishing, as shown in Figure 4.41. (Note that this system may be unstable.) Thus increased

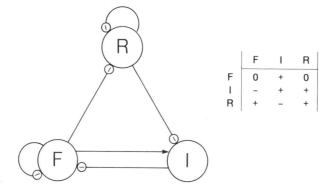

Fig. 4.41

regulatory activity can in fact protect fish. Next, industry, I, sets up a lobby, L, whose activity is aimed at reducing regulation, R. The lobby gets active when R is high, as shown in Figure 4.42.

The lobby protects industry from the intervention of the regulatory agency, increasing I and reducing R. The controversy may continue, with various parties adding variables to the system in different ways. We look at another example of this kind of intervention in the next chapter.

3. Remove a link. This procedure is especially common in physiological research where there is no practical way of physically removing a variable. An inhibitor of particular receptor sites, however, can make a link inactive.

In the system of Figure 4.43, A activates B directly but inhibits B indirectly by way of C. It may be that the direct pathway is stronger and usually overwhelms the indirect path. But if we inhibit the sensitivity of B to A, then the alternative path prevails; input to A reduces B and the negative pathway is confirmed.

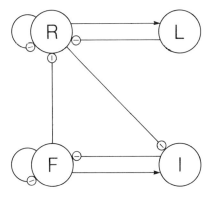 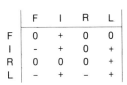

Fig. 4.42

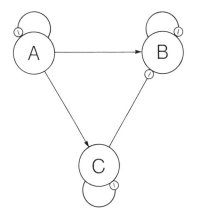

Fig. 4.43

4. Links can be altered. Simple interventions can change the structure of the signed-digraph. The removal of a constant quantity of a variable introduces a positive feedback. For instance, let

$$\frac{dX}{dt} = Xf(X, Y, Z, \ldots) \quad (4.4)$$

Then at equilibrium, $f(X^*, Y^*, Z^*, \ldots) = 0$, or $X^* = 0$, which is the trivial case and we ignore it. Self-damping is calculated from Equation (4.5a)

$$a_{XX} = \frac{\partial}{\partial X}\left(\frac{dX}{dt}\right)\bigg|_* \quad (4.5a)$$

which from (4.4) becomes

$$a_{XX} = \left[X\frac{\partial f}{\partial X}\right]\bigg|_* = X^* \frac{\partial f^*}{\partial X} \quad (4.5b)$$

The sign of a_{XX} depends on whether the function f is increasing or decreasing as a function of X at the equilibrium value: decreasing means there is self-damping and increasing means a positive self-effect.

Suppose we remove now a constant harvest, H, of X, independent of the level of X. Then

$$\frac{dX}{dt} = Xf(X, Y, Z, \cdots) - H \qquad (4.6)$$

At equilibrium X^* is no longer zero but

$$X^* = \frac{H}{f(X, Y, Z, \cdots)} \qquad (4.7)$$

Now, we reevaluate Equation 4.5a and obtain a new value for a_{XX}:

$$a_{XX} = \left[X \frac{\partial f}{\partial X} \right]_* + f(X, Y, Z, \cdots)\bigg|_* = X^* \frac{\partial f^*}{\partial X} + f(X^*, Y^*, Z^*, \cdots) \qquad (4.8)$$

where $f(X^*, Y^*, Z^*, \cdots)$ is considered to be positive for cases of interest here. Thus,

$$a_{XX} = X^* \frac{\partial f^*}{\partial X} + \frac{H^*}{X^*} \qquad (4.9)$$

Even when the derivative of f is negative, the second term of Equation 4.9 is positive, and if large enough can give X a positive self-loop. If X does not appear at all in the function, hence $f = f(Y, Z, \cdots)$, then this feedback is always positive.

Similarly, the addition of a fixed number to X creates a negative feedback. Both procedures alter the table of predictions. For example, suppose that we do not know whether there is a direct link from A to C in Figure 4.44.

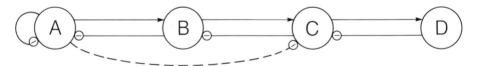

Fig. 4.44

Since the path from A to C leaves both B and D as isolated subsystems, with zero feedback, continuing the path either toward B or D would leave a zero complement. Therefore that link has no detectable effect on the table of outcomes. Since the link runs from A to C, any alterations we make at either of these variables would be irrelevant to the impact of that path. We could introduce self-damping at D by adding a constant amount of D per unit time. Then the path [ACB] would result in a positive effect on B: input to A would increase B. This already happens, however, through the direct path [AB], so that it would not be helpful. The removal of constant quantities of D would place a positive feedback. We then would have the graph shown in Figure 4.45.

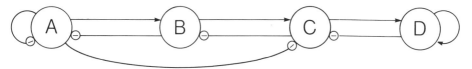

Fig. 4.45

The effect of input to A on B will be positive by the direct path, and negative by way of C. If the self-excitation at D is strong enough, the [AC] link would predominate and the effect would be negative.

An alternative strategy would be to introduce self-damping at B. This would be the complement of the [ACD] path, so that strong enough negative feedback at B would allow us to detect the [AC] link by the negative effect of input to A on D.

In principle we could establish a link between any two variables. If we measure some Y and then add or subtract X at a rate that depends on Y,

$$\text{new } \frac{dX}{dt} = \text{old } \frac{dX}{dt} + g(Y)$$

Then the link from Y to X has the sign of $\partial g(Y)/\partial Y$. This intervention also may introduce self-damping to X. If the old

$$\frac{dX}{dt} = Xf(Y, Z)$$

the change gives a self-damping with the opposite sign from $g(Y)$. But if instead we use

$$\frac{dX}{dt} = X[f(Y, Z) + g(Y)]$$

then X remains without a self-loop. Of course, not all such manipulations will be feasible technically. Therefore it is desirable to consider the uses of many such graph modifications.

Suppose that we know the basic trophic structure of the N, A_1, A_2, H system but suspect the possibility that A_2 inhibits A_1 directly or that H

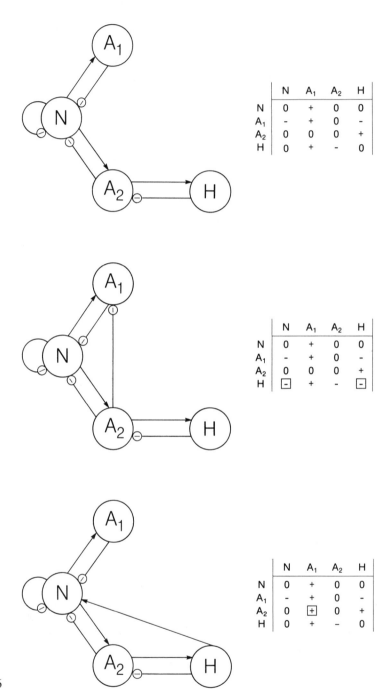

Fig. 4.46

QUALITATIVE PREDICTIONS 115

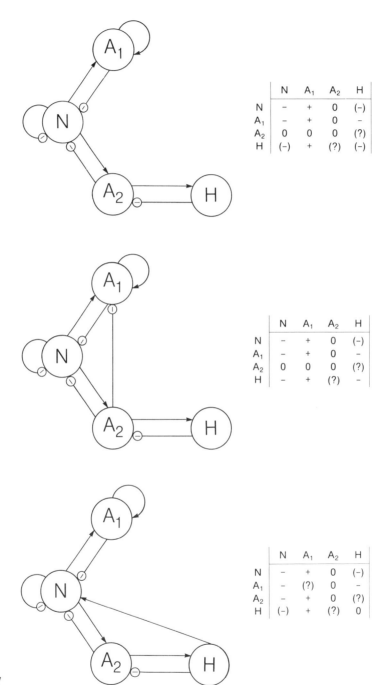

Fig. 4.47

116 QUALITATIVE PREDICTIONS

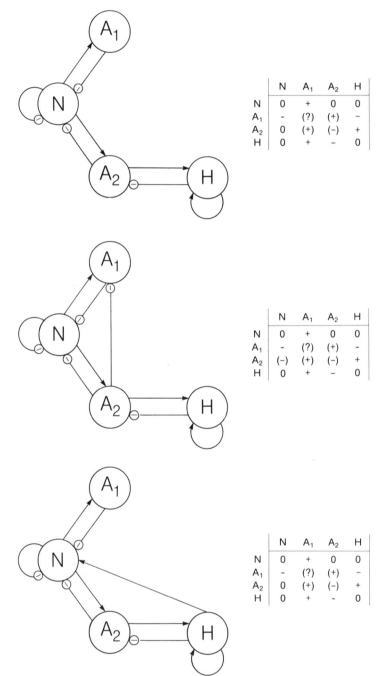

Fig. 4.48

recycles nutrient to N. The three graphs and their tables of predictions are illustrated in Figure 4.46.

These large boxed observations would serve to differentiate among the models. Further confirmation, however, may be desired. Or it may be that for technical reasons we are unable to manipulate H and therefore cannot distinguish the first from the second table. It may even be that we can only assay for N and therefore can only read the first column of the table.

One experimental procedure would be to remove a fixed quantity of A_1 so that A_1 now has a self-excitation loop. The new graphs and their tables are given in Figure 4.47.

The tables have been changed, of course, as indicated by the items in parentheses. But the changes do not permit the separation of the first and second graphs at all. Since the link which is being tested runs from A_2 to A_1, experimental introduction of self-loops to either one will not change either a path or a complement of that link. The third graph now can be distinguished by the negative change in N^* due to an input to A_2.

Suppose instead that we impose a self-excitation loop onto H. This clearly cannot pick up a path from or to H, and therefore will not differentiate the first from the third graphs. The graphs and tables are now as seen in Figure 4.48.

This experiment again gives us the critical observation-impact of a parameter change on N in A_2. The two experiments together would allow a clear differentiation by assaying for changes in N. A negative self-damping loop on H would result in an input to A_2, increasing N in the second graph.

Since the links we are searching for involve all variables in the system, no single change in a self-loop can separate all the graphs. Therefore we might think it desirable to add self-excitation loops to both A_1 and H, simultaneously. The result of such an attempt is shown in Figure 4.49.

The experiment obviously has lost all resolving power: changes in the graph are not additive.

A postulated link may not be detectable for several reasons:

1. Its complements are of zero feedback. (We would have to introduce positive or negative feedback into the complement.)

2. An alternative path has an opposite effect and can outweigh it. (We would have to nullify that alternative path by giving it a zero feedback complement.)

3. The alternative path has the same sign as the hypothetical path; it could mask the effect of our path. (The appropriate strategy would be either to nullify it with a zero complement or reverse the effect of either path with a positive complement.)

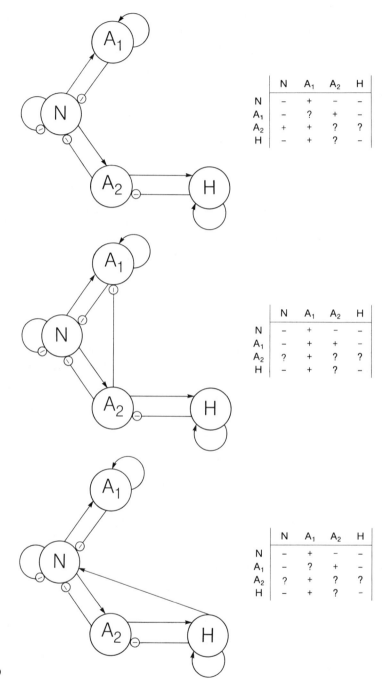

Fig. 4.49

5 From Concept to Model

As different as each model-building procedure is from another there are, nevertheless, shared general characteristics for the model-building process. What we offer here are suggestions of how to build models, including things to do and avoid. Good descriptions and representative mathematical equations of the system of interest are used to produce qualitative models.

Primarily, this is a chapter of examples. We present models that people have proposed and show some consequences of them. We also show how it is possible to use data if they are complete—in sensu, a qualitative analysis table of predictions like those described in Chapter 3—to help build a model; this is called the *inverse problem*.

A good description of the system to be modeled should be obtained. When in doubt—because the reports in the literature disagree, information is unavailable, or a new idea is being tested—then follow all plausible alternatives. Identify the important variables, direct links among them, and interesting phenomena observed. Use previous experience in this or related systems, make some guesses as to possible connections between variables, underlying mechanisms of causation, and possible entry points of perturbations. Active, conscious choices must be made. Important questions to ask are: What is it about the system that makes it interesting? What is uninteresting or of marginal interest? Are there variables that seem to be interacting only with one other variable? How? Is this a satellite variable? What observations are associated with impacts to the system?

Difficulties will arise. For example, in many public health considerations there are parameters that are themselves not influenced by health outcomes. Childhood mortality will be dependent on unemployment rates, but itself does not affect unemployment. Where reciprocal interactions are absent the qualitative signed-digraph models do not add very much to our understanding. In cases where there are a lot of factors feeding into the system of interest but the system is not affected by them, at present these systems are better studied by statistical analysis than by loop analysis. Or, when everything is connected to everything else (or nearly so), then the loop diagram will be uninformative and most predictions ambiguous. Happily, this is unlikely in most cases (Holling, 1978). We can expect the most powerful predictions from loop analysis when there is an intermediate level of complexity.

To illustrate the model-building process, we will discuss a model of diarrhea and malnutrition.

A major cause of child mortality in the Third World is diarrhea. The severity of the bout of diarrhea (and therefore the danger of death) is increased by malnutrition, which lowers resistance to the infective organism. The diarrhea itself increases malnutrition level by loss of appetite, poor absorption of food, and, if accompanied by fever, by a more rapid metabolic expenditure of nutrients.

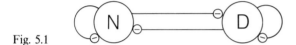

Fig. 5.1

The choice of variables comes from our already existing knowledge. We are interested in the relation between diarrhea and nutrition, so these become variables of the model. It is also known that viral and bacterial agents produce diarrhea and we assume in this model that they are present all the time. Both variables (nutritional status, N, and severity of diarrhea, D) are self-damped: the self-damping of the nutrients reflects their turnover rate in the body (days for B vitamins, perhaps weeks for stored calories and antibody globulins), while the self-damping of D is the recovery rate, reciprocal of the usual duration of a diarrhea bout. The system may therefore be represented by the graph in Figure 5.1. The conditions for stability are:

$$F_1 = -a_{NN} - a_{DD} < 0$$
$$F_2 = (-1)(-a_{NN})(-a_{DD}) + (a_{DN}a_{ND})$$

where F_2 is of unknown sign. If the positive feedback loop at level 2, $a_{DN}a_{ND}$, is weaker than the product of the self-damping terms, the system is stable, leaving the child at some equilibrium level of nutrition and health. Then the problem is to increase the nutritional level, N, and reduce the diarrhea, D. But if $|a_{DN}a_{ND}| > |a_{NN}a_{DD}|$, then the system is unstable. Malnutrition exacerbates diarrhea, which produces more malnutrition in a vicious circle leading eventually to death. Therefore the health-care task becomes one of breaking the vicious circle by reducing the strength of the positive feedback and increasing the self-damping of D.

EXAMPLE 5.1

Demonstrate that the use of malnutrition (MAL) instead of nutrition (N) as a variable will not change the conclusion of the diarrhea–nutrition model.

Solution

Malnutrition is increased by diarrhea while enhancing the malady. The model is illustrated in Figure 5.2. As in the diarrhea–nutrition model, the

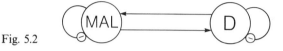

Fig. 5.2

stability is determined by the relative weights of the self-damping terms to the interaction links between D and MAL.

In any model one usually makes some arbitrary choice of variables, a choice that might just depend on idiosyncratic views of the world. □

EXAMPLE 5.2

Suppose in the diarrhea–nutrition model it was uncertain whether microorganisms should be part of the model. Show what their inclusion does in comparison to the conclusions reached in the original model.

Solution

The infective agent, M, one of several kinds of bacteria, appears as an input to D. If the infection rate depends on the microbial population for the area as a whole, an individual contributes a negligible amount to this population. Then there is no link from D to M; M is a parameter of the system and the model is shown in Figure 5.3. But if infection comes from the microorganisms in the

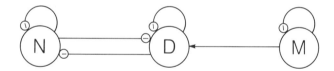

Fig. 5.3

same household (the bacteria that reproduced during the previous bout), then there will be a positive link from D to M. Next we have to decide about self-damping. Although bacteria are self-reproducing organisms, the variable M consists of the bacteria derived from the patient but waiting in the surrounding environment. This variable is not self-reproducing; it may change according to the relation

$$\frac{dB}{dt} = aD - mM$$

where a is the rate at which a sick person adds bacteria to the surroundings and m is a death rate of bacteria. The model now would be that of Figure 5.4.

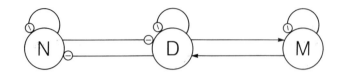

Fig. 5.4

We see that there is another positive feedback at Level 2. There are now two positive feedbacks at Level 2, but three pairs of negative loops of length 1. Further, we now have to consider Level 3, which depends on the negative product of the self-damping terms versus the positive contribution from loops of length 2 times disjunct loops of length 1. The algebra is more complicated but the qualitative result is similar to the original model: a weakening of the positive feedback of length 2 could stabilize the system. A new feature is the ambiguous role of the self-damping

$$F_3 = -a_{NN}a_{DD}a_{MM} + a_{NN}a_{DM}a_{MD} + a_{MM}a_{ND}a_{DN}$$

Therefore if the subsystem of N and D is unstable, a_{MM} enhances the instability, or if the $[D\ M]$ subsystem is unstable a_{NN} enhances the instability.

□

The most frequent loop model construction error made is the omission of links, usually self-damping, but sometimes intermediate links. One way to avoid this omission is to write general, amorphous equations which characterize the relation between the variables.

EXAMPLE 5.3

In the model for two competitors for a single resource one of the competitors has a constant immigration rate and the other is a locally generated stock. Show that the former is self-damped while the latter is not.

Solution

The locally reproducing competitor has a rate of reproduction proportional to its own abundance and the amount of available resource

$$\frac{dH_1}{dt} = H_1 g(R)$$

At equilibrium

$$\frac{dH_1}{dt} = 0 = H_i^* g(R^*)$$

so $g(R^*) = 0$. Then

$$a_{H_1 H_1} = \frac{\partial}{\partial H_1}\left(\frac{dH_1}{dt}\right) = g(R)\bigg|_{R^*} = 0$$

The second competitor has a growth rate based on two parts: one consists of a rate of reproduction proportional to its own abundance and the resource level, as is the case with H_1; the other part is the constant immigration rate.

$$\frac{dH_2}{dt} = H_2 f(R) + m$$

At equilibrium

$$\frac{dH_2}{dt} = 0 = H_2^* f(R^*) + m$$

or

$$f(R^*) = \frac{-m}{H_2^*} < 0$$

Then

$$a_{H_2 H_2} = \frac{\partial}{\partial H_2}\left(\frac{dH_2}{dt}\right) = f(R)\bigg|_{R^*} < 0$$

The model is shown in Figure 5.5. □

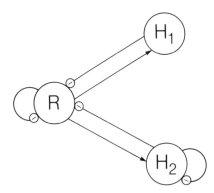

Fig. 5.5

Qualitative models are built to answer questions about an event or activity to better understand or explain the observations. Variables of interest are partially decided by the reasons for building a model and partially by the system structure. For example, a study of a widely fluctuating predator density will obviously include the predator as a variable, but if the observation is made with intent to use a herbicide then the agency decision rule for the application becomes a potential variable; in turn, the system-defined links of these variables to other variables now becomes incorporated into the model—the activities of various and opposite lobbying groups, the target species, its food web, and that of the predator. If the observations of predator fluctuation are brought on by a concern for associated outbreaks of parasitism, then a different set of variables such as the vector of alternative hosts, and corresponding food webs come into play. There will be many variables that overlap in the two scenarios just described but the different variables reflect that the problem under study identifies the system boundaries.

In some cases links between variables may be either unknown or indeed absent, but possible to build into the system to alter its dynamics. In a later section we discuss this latter possibility in more detail. For the present we focus on a model of migraine that might be based on blood circulation in the head, with consideration of how factors like stress or food or alcohol are agents producing or enhancing migraine.

During migraine the cerebral blood flow increases in the lateral side of the head—which is important only because it is associated with a loss in autoregulation of the cerebral blood circulation (Spierings, 1980; Rose and Gawel, 1979). Consequently, suppose intensity of migraine comes about through the degree of loss in cerebral blood regulation, but as inferred by an increase in cranial blood flow. We take the amount of pain associated with migraine (M) to indicate the intensity of an attack. The model is depicted in Figure 5.6. This model is structurally identical, from a consideration of stability criteria, with the model of nutrition and diarrhea of Figure 5.1.

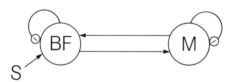

Fig. 5.6 BF = amount of cranial blood flow, M = intensity of migraine, S = amount of stress.

Either through stimulation of adrenaline or noradrenaline production, stress seems implicated in the onset of migraine, although this is unknown. Suppose stress causes a loss of cranial blood regulation—hence increased blood flow. The system is being driven by the blood flow end (BF) and an increase in BF will be associated with an increase in migraine (M), provided the system is stable.

When feedback at level two produced by the interaction between cranial blood flow and migraine intensity is stronger than the product of the self-damping links between the two variables, then the system is unstable. The strength of the self-damped links will vary among individuals and may produce opposite response to the same inputs: reports, for example, that exercise in some cases brings on a bout of migraine, while in other cases it seems to prevent or prolong the periods between migraine. We do not have a theory of migraine but we suggest that an examination of the difference between the strengths of positive and negative feedbacks may reveal the relation between what appear as randomly occurring events or inconsistent outcomes from application of the same stimuli.

Now suppose that blood flow also is affected by food (F), by alcohol (A), and other factors like that. If there is no feedback from BF to F, to S, or to A, and there is also no feedback from the migraine to those variables, then this is the simple system of Figure 5.7, which has a large number of inputs. A large number of inputs is not very interesting, and the model is not of much benefit. There is only a minimum amount of information coming from the loop of length 2.

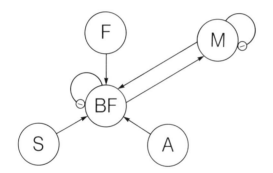

Fig. 5.7

The model of Figure 5.7, posed in this way, means a legitimate conclusion is to provide the patient with drugs to reduce cranial blood flow as a way to control migraines.

There is, however, the option to look far afield: looking at the stress level, S; at the quality of food, F; at alcohol, A (see Rose and Gawel, 1979). This, however, is not a good problem for loop analysis because most of the relevant variables are not structured with reciprocal feedbacks.

Loop analysis becomes important when there is reciprocal interaction among the variables. The cranial blood flow–migraine case should be analyzed by loop analysis; yet the role of the other variables is best analyzed with ordinary statistical methods.

One suspicion might be that the incidence of migraine is not solely related to cranial blood flow. For example, if the perception of migraine is itself an anxiety, there may be a pathway from migraine to stress. A link is missing that was unknown. A positive feedback exists from *S* to *BF* to *M*, as shown in Figure 5.8. This positive feedback signifies that when the migraine is set off for

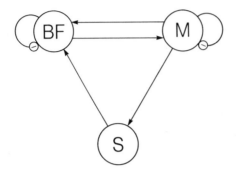

Fig. 5.8

any reason it may be kept up. It may become migrainous as a phenomenon. Suppose migraine affects food habits, perhaps through a knowledge of health foods, or a knowledge of alcohol as a blood-vessel dilator can be used to control or limit intake with a migraine-prone individual. Building these links can change the model (Figure 5.9).

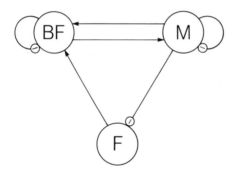

Fig. 5.9

Now feedback at Level 3 is negative and the person is not exacerbating a possibly worse situation. In addition, the alcohol limitation also may be a factor in increasing the strength of the self-damping loops.

Part of a health care program might include trying to build links. In practice, there is not going to be a national economic program based on migraines. So the effect, say, of unemployment on provoking migraine has to be ignored. Employment is exogenous. Conversely, if greatly prolonged periods of stress on a job lead to repeated outbreaks of migraine and missed days of work, then that might be a strong influence to build a stress reduction program into the job. These kinds of considerations grow out of the analysis of simple models.

In short, starting with a naive model which includes only direct causes of action leads to very little new information. But a willingness to suspect new links or to build them in can lead to new questions and ideas.

Ecological Examples

A typical case of model building starts with a fairly good description of the system. Briand and McCauley (1978) have described lake planktonic communities with a loop model as shown in Figure 5.10.

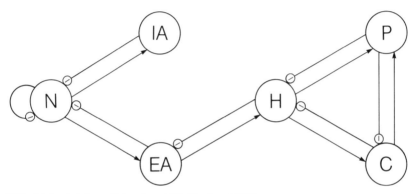

Fig. 5.10 (Adapted from Briand and McCauley, 1978)

The six variables of interest are: nutrients (N), inedible algae (IA), edible algae (EA), herbivores (H), carnivores (C), and planktivores (P). According to Briand and McCauley, edible and inedible algae categories are due to the size, taste, and morphology selection of grazers. The grazers are eaten by carnivores, and both of these are consumed by planktivorous predators. Only the nutrients are self-damped in this system. The model structure, the links among the six variables, determines the predicted consequences on equilibrium levels due to changes in the growth rates of any of the six variables.

A table of predictions for the change in the equilibrium abundance of the six variables due to changes in the growth rate parameters in any one of them has been given by Briand and McCauley, outlined in Table 5.1.

The nutrient level is unaffected by any parameter input except that entering through inedible algae (IA). An increase in nutrients has no effect on any other variable except inedible algae. Along the diagonal of Table 5.1 are zeros: Any change in a parameter that increases the growth rate of a variable

Table 5.1 Table of predictions (adapted from Briand and McCauley, 1978)

	N	IA	EA	H	C	P
N	0	+	0	0	0	0
IA	−	0*	0**	−	+	−
EA	0	0**	0**	+	−	+
H	0	+	−	0	0	0
C	0	−	+	0	0	+
P	0	+	−	0	−	0

* The original table is incorrect. This zero should be +.

** The authors seem to assume the [HPC] subsystem has zero feedback.

produces no change in its equilibrium value. This cannot be true and is in fact an error in the table. We showed in Chapter 4 that for a stable system any row or column cannot have all zeros because this would mean the overall feedback is zero and would violate the criteria for stability. The same results apply here—for all the variables of a system to be immune to change in their own abundance when there is a change in their own parameters means the overall feedback of the system is zero. Change entering through the algae alters the abundances of all three consumers. Likewise, change entering through the consumers shifts the algae abundances, primarily, with the single exception that positive inputs to the carnivores (C) do increase the planktivorous predators (P).

Briand and McCauley identify "cultural" eutrophication and experimental nutrient enrichment as showing increases in inedible algae with no change in N, as is predicted by the model. Attempts to eliminate predatory fish (C) showed zero change in herbivore (H) standing crop, although species composition shifted. By suspending polyethylene tubes in Lake Heney, Ottawa, Briand and McCauley were able to reduce herbivore and carnivore levels and found a significant increase in inedible algae with no change in either the herbivore or carnivore level. This agrees with the row predictions of N and H found in Table 5.1. (Note: Reverse the signs because the experiment is for a change that decreases the parameters of H and C).

It has been suggested (Finerty, 1980) that some mammalian populations go through regular cycles of abundance and scarcity over a period of several years. Explanations have been offered in terms of predatory–prey interactions (Hutchinson, 1978): That predators of the snowshoe hare (*Lepus americanus*) in central Alberta probably exacerbate amplitude fluctuations, reduce the cycle frequency, and cause synchronous cycles in neighboring hare populations, within the bounds of mobility of the predator (Finerty, 1980); or, that the lemming (*Lemmus* spp.) exhibits fluctuations in concert with its predators as a special form of predator avoidance (Leigh, 1975).

Finerty (1980) gives another account of cycles and we follow his example here.

For a particular model of small mammals such as hares or microtine rodents (voles, lemmings, mice), the notion of a survival habitat and colonizing habitat plays an important role in the explanation of cycles. The amount of edible vegetation in a survival habitat, S, and colonizing habitat or dispersal sink, C, for a resident small herbivorous mammal population in each habitat, H and $\tilde{H}$, respectively, has been modeled as in Figure 5.11 (modified from Finerty, 1980). The vegetation in both the survival and colonizing habitat is self-damped, from the general rule that this is true if primary producers or the lowest elements of the food web are included in the model. The mammal population in the colonizing habitat is self-damped due to immigration from the survival habitat.

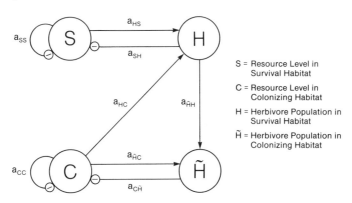

Fig. 5.11 (Adapted from Finerty, 1980)

Finerty also postulates that a good plant yield in the colonizing habitat encourages emigration of mammals from the survival habitat. It is presumed that mammals in the colonizing habitat do not return to the survival habitat—therefore the name dispersal sink. The first stability criterion $F_i < 0$, for $i = 1$ to 4, is met since all loops are negative. Instability comes about from a failure of the system to meet the second stability criterion.

The second criterion is a weighing of the relative strengths of long loops to short loops. The loop of length 3 in the model shown here, $a_{HC}a_{\tilde{H}H}a_{C\tilde{H}}$, is counterbalanced by the loops of lengths 1 and 2. The product of the lower-order loops will have terms that cancel with disjunct loops at the highest level, leaving the conjunct loops of lower order against the long loops from the highest-order feedback. Thus the second criterion for stability is that

$$\{F_1F_2\} + \{F_3\} > 0$$

which in this example reduces to

$$+ (a_{SS}^2 a_{CC} + a_{SS} a_{CC}^2 + a_{SS} a_{HS} a_{SH} + a_{CC} a_{\tilde{H}C} a_{C\tilde{H}}) - (a_{H\tilde{H}} a_{HC} a_{C\tilde{H}}) > 0$$

The expression is positive when the first parenthetical term is stronger than the second; however, if $|a_{H\tilde{H}} a_{HC} a_{C\tilde{H}}|$ is larger than the first term, an oscillation will ensue (see Finerty, 1980). In this case there is only one expression for the second criterion. If there are no loops of length 5 or greater, then the next expression which includes the length 4 loops does not have to be evaluated; it is satisfied if the previous level expression is. (See the discussion of the Lienard–Chipart theorem in Chapter 6.)

It seems true that dispersal is important for both hares and lemmings from survival habitats, and that this model has at least one aspect in correspondence with nature. Finerty goes on to show that neither self-damping of the herbivore population nor addition of a predator in the colonizing habitat alters the main conclusion: With increasing strength of the long negative feedback, $[CH\tilde{H}]$, the existence and influence of the colonizing habitat to the mammal population, the more likely the population will exhibit cycles.

EXAMPLE 5.4

Demonstrate that a lower reproductive response to crowding by voles in the survival habitat will not change the conditions for population oscillations of the model of Figure 5.11.

Solution

If we assume that voles have lower reproductive output under crowded conditions (see Tamarin, 1978), then the small mammal population in the survival habitat also becomes self-damped, as shown in the revised model of Figure 5.12. Crowding has produced self-damping on the survival habitat vole population. We have included self-damping on the colonization habitat population since H is not self-reproducing but receives migrants.

Denoting feedback at Level 3 for this modified model as G_3, then

$$G_3 = F_3 - a_{HH}(a_{SS}a_{CC} + a_{SS}a_{\tilde{H}\tilde{H}} + a_{CC}a_{\tilde{H}\tilde{H}}) - a_{HH}(a_{C\tilde{H}}a_{\tilde{H}C})$$

The new self-damping loops make stability more likely. The source of possible oscillation is still the $[CH\tilde{H}]$ loop. □

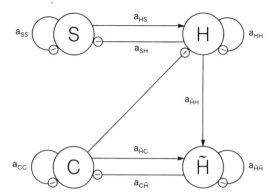

Fig. 5.12

EXAMPLE 5.5

The management of a bay scallop (*Argopectin irradians irradians*) fishery can consist either by control over the fishing industry or implementing a protection procedure to increase the winter survival rate of juveniles. Bay scallops are edible invertebrate bivalves that live for two years and obtain food by filtering materials from the water column. Frequently the young seed scallops (<1 year old) are prohibited from the catch harvest (as on the islands of Nantucket and Martha's Vineyard, Massachusetts). For an abundant adult population, with a high price-per-pound of "eye" (the edible scallop abductor muscle) there will be many fishing boats dredging for scallops. Later in the season, with adult scallop density down, the number of vessels declines. A drop in the price of the bay scallop is followed by a decline in the fishing industry. What difference would there be in a local town regulatory commission empowered to act directly on the fishing industry by control of the number of fishing vessels or price of fishing licenses, versus a commission set up to take action to increase the number of juvenile scallops that survive a winter season? One proposed method is to collect the juveniles washed ashore by winter storms and return them as soon as possible to the bay (Kelley, 1982). (See the discussion of fisheries management in Chapter 4.)

Solution

Two models of scallop management are illustrated in Figure 5.13. Both models show juvenile scallops as the supply of adult scallops, and the adults as the source of the local population of juveniles, leading to positive links between them and the occurrence of self-damping. The fishing industry is self-damped because the number of trawlers joining the industry is based on the price of licenses, population size of scallops, and access to markets, but not the size of the present industry. Self-damping on the price of the scallop meat

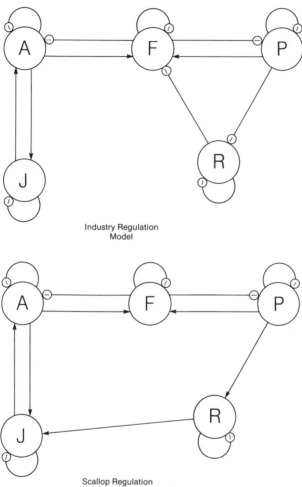

Fig. 5.13

comes about by the market. The regulatory commission in the industry regulated model is self-damped because it can act only within its local mandate for enforcement and is bounded by state and federal laws, budget, personnel, bureaucratic lag-time, and is affected by lobbying pressure. As the price of the scallop on the open market drops, the regulatory commission responds to pressure for control on the amount of access to scallops.

In the scallop regulation model, the regulatory commission intervenes in the survival of juvenile scallops provided it has sufficient staff, equipment, and money; and responds to price by going into action when scallop price is high. It becomes unprofitable and politically unrealistic for a shellfish authority to maintain expensive biological intervention when the net income or number of jobs supported by the industry is small, as would be the case for a drop in scallop price.

Positive loops appear in feedback levels greater than Level 1 of the industry regulation model. At Level 2*

$$F_2 = (-1)\left[+L_A(L_J + L_P + L_R) + L_J(L_F + L_P + L_R) \right. $$
$$\left. + L_F(L_P + L_R) + L_P L_R \right] - L_{AF} - L_{FP} + L_{AJ}$$

Clearly positive feedback is indicated in the two loops of the juvenile–adult scallop subsystem. In a linear system (or subsystem), which is entirely closed and has no inputs from the outside, the overall feedback is zero. We can see this by writing the equations of the juvenile–adult subsystem:

$$\frac{dA}{dt} = rJ - mA - fAI$$

$$\frac{dJ}{dt} = sA - dJ$$

where r = recruitment rate of juveniles into adult population
       m = natural mortality rate of adults
       f = mortality rate of scallops per unit fishing effort
       s = fecundity rate of adults
       d = natural mortality rate of juveniles

---

* The correspondence between the L notation and that previously used should be obvious. Consider these examples:

$$-[A] = -a_{AA} = -L_A$$
$$[A] = +a_{AA} = +L_A$$
$$[AB] = (a_{AB}a_{BA}) = L_{AB}$$
$$-[AB] = -(a_{AB}a_{BA}) = -L_{AB}$$
$$-[A][B] = -(a_{AA})(a_{BB}) = -L_A L_B$$
$$[ABC] = (a_{BA}a_{CB}a_{AC}) = L_{ABC}$$

Note that in both the bracket notation $[IJK\cdots]$ and the $L_{ijk}\cdots$ notation the symbols are read "from i to j to k to $\cdots$" which is the opposite from the $a_{ji}a_{kj}a_{..}\cdots_{k.}$. If only the second subscripts are read on the link symbols, then the two notations do read the same.

Then

$$a_{AJ} = \frac{\partial}{\partial J}\left(\frac{dA}{dt}\right) = r \qquad a_{AA} = \frac{\partial}{\partial A}\left(\frac{dA}{dt}\right) = -(m + fI)_{I*}$$

$$a_{JA} = \frac{\partial}{\partial A}\left(\frac{dJ}{dt}\right) = s \qquad a_{JJ} = \frac{\partial}{\partial J}\left(\frac{dJ}{dt}\right) = -d$$

At equilibrium

$$0 = rJ^* - mA^* - fAI^* \quad \text{and} \quad 0 = sA^* - dJ^*$$

So

$$(m + fI^*) = r\frac{J^*}{A^*} \quad \text{and} \quad J^* = s\frac{A^*}{d}$$

Finally

$$L_{AJ} = a_{AJ}a_{JA} = +rs$$

$$L_A L_J = (a_{AA})(a_{JJ}) = \left(\frac{-rs}{d}\right)(-d) = +rs$$

The feedback of the closed subsystem between juvenile and adult scallops, $L_{AJ} + (-1)L_A L_J$, is zero.

The same positive loop between juvenile and adult scallops causes positive feedback at Level 3.

$$\begin{aligned} F_3 = &-L_A L_J(L_P + L_R + L_F) - L_A L_P(L_R + L_F) - L_A L_R L_F \\ &- L_J L_P(L_R + L_F) - L_J L_R L_F - L_P L_R L_F - L_A L_{FP} \\ &- L_J(L_{FA} + L_{FP}) - L_P(L_{FA} - L_{JA}) \\ &- L_R(L_{FA} + L_{FP} - L_{JA}) + L_F L_{JA} - L_{FPR} \end{aligned}$$

Rearranging terms as shown

$$\begin{aligned} F_3 = &-\{L_A L_P(L_R + L_F) + L_A L_R L_F + L_J L_P(L_R + L_F) \\ &+ L_J L_R L_F + L_P L_R L_F + L_A L_{FP} + L_J(L_{FA} + L_{FP}) + L_P L_{FA} \\ &+ L_R(L_{FA} + L_{FP}) + L_{FR}\} + (L_P + L_C + L_F)(L_{JA} - L_A L_J) \end{aligned}$$

brings out where the potential for positive feedback occurs. Again, however, the positive term drops out.

The scallop regulation model has the same positive Level 2 feedback loop as the fishing regulation model balanced by Level 2 self-damping on juvenile and adult scallops.

In the industry regulation model parameter changes in any of the variables except the scallops have no effect on the fishing industry since all pathways leave the scallops as an isolated subsystem with zero feedback. Consequently, the graph structure shows that a commission set up to regulate this scallop fishing industry will be ineffectual. □

Even systems that have been recognized and studied over long periods may have interactions among variables that are either unknown or uncertain. Li and Moyle (1981) describe the pre-1880 pelagic ultraoligotrophic Lake Tahoe as a community consisting of algae (A), cladocerans (Cl), and two fishes, tui chubs (*Gila bicolor pectinifer*) (*Tui*) and Lahontan cutthroat trout (*Salmo clarkii hernshawii*) (*Cut*), shown in Figure 5.14. The interactions are deduced by observing similar systems.

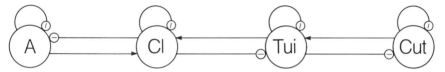

Fig. 5.14

One major change in Lake Tahoe since 1890 has been the introduction of the opossum shrimp (*Mysis relicta*) and three fishes: kokanee (*Oncorhynchus nerka*), lake trout (*Salvelinus namaycosh*), and rainbow trout (*Salmo gaird-*

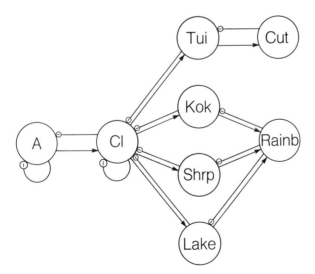

Fig. 5.15 (Adapted from Li and Moyle, 1981)

neri). To determine the direction (signs) of interactions between the current members of Lake Tahoe, Li and Moyle looked at the studies of feeding habits of the Lake Tahoe fishes, the decline of cladocerans and kokanee, and the biology of the opossum shrimp. Consumption of zooplankton by the shrimp and all the fishes except the cutthroat trout indicate competition in the oligotrophic, food-limited Lake Tahoe.

The Li and Moyle model of Figure 5.15 includes all major groups of the pelagic community. The feature which is important to the post-1890 Lake Tahoe ecology, however, is the competition among the introduced higher-order consumers—*Kok*, *Shrp*, and *Lake*.

An analysis of the competition can be obtained from the submodel of Figure 5.16.

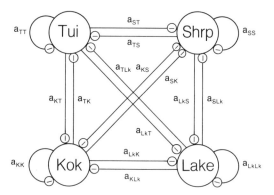

Fig. 5.16

We see in Figure 5.16 that feedback at Level 2 is both positive and negative. The system is unstable unless self-damping outweighs pairwise competitive interactions. Stability is obtained for feedback at Level 3 when the longer, length 3 negative loops are stronger than the product of shorter-length loops, while stability at Level 4 feedback demands that the longer, length 4 positive loops be weaker than the product of disjunct negative feedback loops.

EXAMPLE 5.6

Identify the ways the introduction of species to Lake Tahoe can bring on instability and oscillations to the pelagic community, using the model of Figure 5.16.

Solution
Feedback for all four levels (using the $L_{ij}\cdots$ notation) are

$$F_1 = -[L_{Lk} + L_S + L_R + L_K]$$

$$F_2 = -[L_{Lk}L_S + L_{Lk}L_R + L_{Lk}L_K + L_SL_R + L_SL_K + L_RL_K]$$
$$\quad + [L_{ST} + L_{RT} + L_{KT} + L_{RS} + L_{KS} + L_{KR}]$$

$$F_3 = -[L_{Lk}L_SL_R + L_{Lk}L_SL_K + L_SL_RL_K]$$
$$\quad + [L_{Lk}L_{RS} + L_{Lk}L_{KS} + L_{Lk}L_{KR} + L_SL_{RT} + L_SL_{KT}$$
$$\quad + L_SL_{KR} + L_RL_{ST} + L_RL_{KT} + L_RL_{KS}$$
$$\quad + L_KL_{ST} + L_KL_{RT} + L_KL_{RS}]$$
$$\quad - [L_{STR} + L_{STK} + L_{LkSR} + L_{LkSK} + L_{RTK} + L_{LkRK}$$
$$\quad + L_{RSK} + L_{SRK}]$$

$$F_4 = -[L_{Lk}L_SL_RL_K] - [L_{ST}L_{KR} + L_{RT}L_{KS} + L_{KT}L_{RS}]$$
$$\quad - [L_{Lk}L_{SRK} + L_{Lk}L_{RSK} + L_SL_{RTK} + L_SL_{LkRK} + L_RL_{STK}$$
$$\quad + L_RL_{LkSK} + L_KL_{STR} + L_KL_{LkSR}]$$
$$\quad + [L_{Lk}L_SL_{KR} + L_{Lk}L_RL_{KS} + L_{Lkk}L_{RS}$$
$$\quad + L_SL_RL_{KT} + L_SL_KL_{RT} + L_RL_KL_{ST}]$$
$$\quad + [L_{STRK} + L_{STKR} + L_{LkSRK} + L_{LkSKR} + L_{RTKS} + L_{LkRSK}]$$

Instability will arise in the higher three feedback levels if the pairwise interactions between species outweigh their self-damping products. At Level 4 feedback the onset of instability occurs when the long, length 4 pathways become large. In other words, the stability of the system with the introduced species becomes undermined. Either the introduced species have to be highly density regulated or a high level of ongoing stock recruitment is necessary to ensure a large amount of self-damping to get system stability. □

Changes in the Graph Structure

Experimental (planned) or other types of intervention in a system may not only provide an input to the system but also may change the structure. For example, suppose that a herbivore (H) behaves as

$$\frac{dH}{dt} = H(A_3 - \theta) \tag{5.1}$$

This herbivore consumes several different grains, which can be combined linearly to form the aggregate variable, A_3, just as in the algal example of

Chapter 4. There may be some other combination of the grains which forms a consumer variable A_4 for nutrient, grain, or other resources. There may be yet other combinations for the grains which are consumer or resources for other unspecified variables in this system.

Suppose the herbivore is an economically important organism that is going to have a certain number harvested every year, to attain the meat industry target production goal. Therefore, Equation 5.1 becomes

$$\frac{dH}{dt} = H(A_3 - \theta) - E \qquad (5.2)$$

the original equation minus an economic consumption rate, E. This system has a positive self-loop (self-enhancement) because the derivative of 5.2 with respect to H is

$$\frac{\partial}{\partial H}\left(\frac{dH}{dt}\right) = (A_3 - \theta)$$

Whenever there is an equilibrium, $(A_3 - \theta)$ is positive. The pattern of economic exploitation causes the herbivore to acquire a positive feedback. A change arising in the system at A_3—a positive input to A_3, say—will result in a decrease in A_3. Similarly, a positive input to A_4 will result in a decrease in A_4, and so on.

The self-enhancement on H is also capable of changing the feedback of the whole system, making it unstable; a constant harvesting rate would be a source of instability.

Another strategy for exploitation of the herbivore population is to remove them based on the number of A_4's. A census of A_4 is made (that is, the abundance of the grains is estimated and the linear combination calculated to obtain A_4) following the rule that as A_4 increases, remove more herbivores. This is the equivalent of building a link through the experimentor to the herbivore, H. If the experimentor works rapidly enough, then you can ignore the link for dynamic purposes; a slowly responding experimentor means the loop is longer, such as might be the case in a weed-control process.

A continuous experimental manipulation, not just a single one-step perturbation, can change the structure of the signed digraph, and it may be important to know about this change in graph structure. It may be desirable to change the structure of a graph; that may even be the purpose of the intervention.

In cardiological research this can be the case. Suppose cardiologists are attempting to find out the effect of a particular variable where two variables of a system seem to have the same effect. For example, a neural response coming from the pressure sensors in a vein affecting blood pressure may also be something that acts directly on the kidney. The cardiologists apply one drug which is a nervous inhibitor, thus cutting one of the links to observe unambiguously the effect of the other. Thus, not only is the system perturbed; the structure is also changed.

Systems can be linked that were previously disconnected, using some kind of on-line feedback. In biofeedback a new and more rapid link is established in the system, taking place through the cerebral cortex and electronic equipment.

Qualitative models should help us think about how to make links between variables that might not otherwise exist. The model is both of the actual and the possible. Moreover, the model often can highlight that—through intervention we become part of what we want to control.

From Data to Graph

Suppose first that under controlled conditions we are able to introduce inputs into each variable of a system and observe all the variables until they reach an equilibrium. Then we begin with a table of outcomes for each experiment, and the problem is to reconstruct the graph. There are three ways of doing this. The first is a stepwise construction in which inferences are drawn from particular effects to particular links. The second method is an algebraic inversion of the matrix of the graph, and the third is an enlightened trial and error in which alternative models are produced that make use of any partial knowledge that may be available until a satisfactory fit is achieved.

STEPWISE RECONSTRUCTION

We assume we are able to cause a parameter change—an input—to each variable in turn and allow enough time for the system to reach equilibrium. Then we would have a complete table of predictions. For example, suppose we have the table

Parameter change in (input to)	Change in equilibrium	
	X	Y
X	0	+
Y	−	+

Since the equilibrium level of X does not change with input to X, the complement of X has zero feedback. That is, Y is a satellite. But input to Y

increases Y, so that X is self-damped. The effects of input to Y on X and to X on Y identify the links and allow the whole graph to be reconstructed as Figure 5.17.

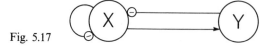

Fig. 5.17

EXAMPLE 5.7

Find the graph for the table

	N	A_1	A_2
N	0	0	+
A_1	−	+	+
A_2	0	−	+

Solution

First, consider the effect of input N on N. Since N^* (the equilibrium value of N) does not increase, there must be some increased removal rate. This can only come from A_2, which has increased. Therefore there is the negative link, shown in Figure 5.18.

Fig. 5.18

Input to A_2 does not increase N, however. Therefore the A_2, N link has zero complement in A_1, which cannot be self-damped. Further, there is no path A_2-A_1-N. But since input to N increases A_2, and a direct N-A_2 link would have a zero complement, there is a positive path N-A_1-A_2.

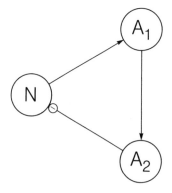

Fig. 5.19

Input to A_2 decreases A_1. It cannot do so directly, since there is no A_2-A_1 link. Therefore A_2-N-A_1 is negative, and since A_2-N is negative, N-A_1 is positive. We now have the graph depicted in Figure 5.19.

The input to A_1 increases A_1. Therefore its complement has negative feedback, giving the graph in Figure 5.20.

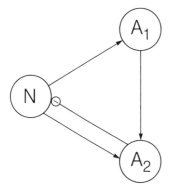

Fig. 5.20

Similarly, input to A_2 increases A_2 so that the complement $[N, A_1]$ has negative feedback, shown in Figure 5.21.

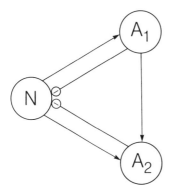

Fig. 5.21

Finally, input to A_1 increases A_2. Since the $[A_1NA_2]$ path has a negative effect it must be outweighed by the direct positive link, hence self-damping on N. The model becomes the one illustrated in Figure 5.22. □

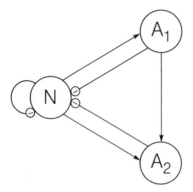

Fig. 5.22

MATRIX INVERSION

The loop algorithm (equation 3.1; see Appendix Equation A.28) is derived from the set of equations for each parameter change, C

$$\sum_j a_{ij} \frac{\partial X_j^*}{\partial C} + \frac{\partial f_i}{\partial C} = 0 \tag{5.3}$$

with one equation for each i. The result is the table of predictions, an $n \times n$ matrix whose entry b_{ij} is the effect of a parameter change to i on the equilibrium level of X_j. The same procedure can be performed backwards, starting with the b matrix, producing the corresponding graph and finding the table of predictions for that graph. The elements of that table, a_{ij}, are the links from X_j to X_i.

EXAMPLE 5.8

Find the graph for the table given below.

		j	
		X_1	X_2
i	X_1	0	+
	X_2	−	+

Solution

Each b_{ij} is a link from X_i to X_j. The table produces the intermediate graph shown in Figure 5.23.

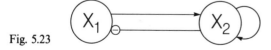

Fig. 5.23

From the intermediate signed-digraph we get the table

$$\begin{array}{c|cc} & X_1 & X_2 \\ \hline X_1 & - & + \\ X_2 & - & 0 \end{array}$$

This table represents each link a_{ij} as

$$j\ \begin{array}{c|cc} & \multicolumn{2}{c}{i} \\ & X_1 & X_2 \\ \hline X_1 & a_{11} & a_{21} \\ X_2 & a_{12} & a_{22} \end{array}$$

The model is then the one in Figure 5.24. It can be verified that this model produces the original table. □

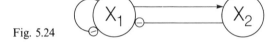

Fig. 5.24

EXAMPLE 5.9

Find the loop model for the table of predictions shown below.

$$\begin{array}{c|ccc} & X_1 & X_2 & X_3 \\ \hline X_1 & + & 0 & 0 \\ X_2 & - & + & + \\ X_3 & - & - & 0 \end{array}$$

Solution

The intermediate signed-digraph is illustrated in Figure 5.25. Note that $F_3 > 0$, so that in obtaining the intermediate table of predictions each path that has a negative complement produces a sign opposite to the path

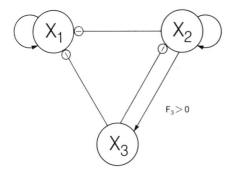

Fig. 5.25

value—just the reverse of what is our usual case. The intermediate table of predictions is

	X_1	X_2	X_3
X_1	−	0	0
X_2	+	0	+
X_3	−	−	−

Thus the loop model for the original table of predictions is written from the intermediate table and is presented in Figure 5.26.

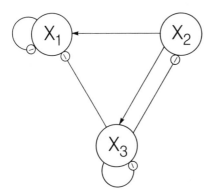

Fig. 5.26

The table of predictions from this model is

	X_1	X_2	X_3
X_1	+	0	0
X_2	?	+	+
X_3	−	−	0

The ambiguity arises in the final table of predictions because input to X_2 has a positive effect by its direct link and a negative effect by way of X_3. Either a plus or minus in this position leads to the same graph. □

EXAMPLE 5.10

Find the loop model corresponding to the table of predictions given here.

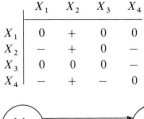

	X_1	X_2	X_3	X_4
X_1	0	+	0	0
X_2	−	+	0	−
X_3	0	0	0	−
X_4	−	+	−	0

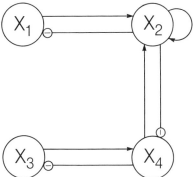

Fig. 5.27

Solution

The intermediate signed-digraph is shown in Figure 5.27. The corresponding intermediate table of predictions is

	X_1	X_2	X_3	X_4
X_1	−	+	+	0
X_2	−	0	0	0
X_3	−	0	0	+
X_4	0	0	−	0

and the final loop model is shown in Figure 5.28. This model produces the table of predictions that we started with. □

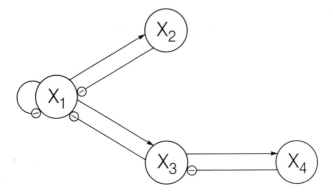

Fig. 5.28

EXAMPLE 5.11

Find the loop model for this table of predictions

	X_1	X_2	X_3
X_1	+	0	+
X_2	0	0	+
X_3	+	−	+

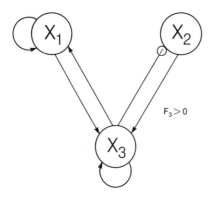

Fig. 5.29

Solution

Figure 5.29 presents the intermediate signed-digraph model. Feedback at Level 3 is positive, $F_3 > 0$. The intermediate table of predictions is

	X_1	X_2	X_3
X_1	−	+	0
X_2	−	?	+
X_3	0	−	0

The final loop model can be drawn as that shown in Figure 5.30.

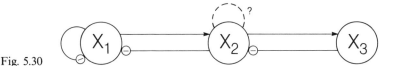

Fig. 5.30

There is one ambiguity in the last signed-digraph. This arises because the unknown a_{22} in the intermediate table of predictions affects the final table of predictions only in the complement of X_3, as $a_{12}a_{21}-a_{11}a_{22}$. This term must be negative, but a_{22} may be positive if it is not too large. □

Sometimes a complete experiment cannot be performed or all observations cannot be made. Still, we can use the same procedure to get some information. For example, suppose we have the table

	X_1	X_2	X_3
X_1	0	0	+
X_2	a	b	c
X_3	0	−	+

where a, b, and c are unknown.

The intermediate graph is given in Figure 5.31.

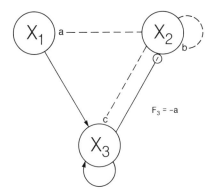

Fig. 5.31

The sign of the overall feedback $F_3 = -a$. We get the intermediate table of predictions

	X_1	X_2	X_3
X_1	?	$-a$	$-ab$
X_2	$-$	0	$+$
X_3	$-$	0	0

The corresponding loop model is presented in Figure 5.32.

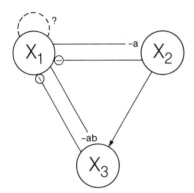

Fig. 5.32

If we assume, however, that the system is stable, we can go further: X_1 must be self-damped, and the loop $[X_1 X_2 X_3]$ must be negative so that a is negative. Then we have as our model Figure 5.33.

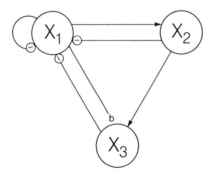

Fig. 5.33

We are left with an ambiguous link, the sign of which is that of the missing observation, b. We have learned that the missing observation a must be negative if the system is stable, and that the missing observation c does not affect reconstruction of the signed digraph.

As the system gets bigger, and more and more observations are missing, this procedure becomes less useful.

ENLIGHTENED TRIAL AND ERROR

This method presumes that we have some previous knowledge of the system from which we can postulate a rough skeleton of the graph. In ecosystems this skeleton usually will be trophic relations. In addition to the skeleton of the system there may be all sorts of other links: bryozoans provide substrate for the settling of scallop larvae; kelp may hide crabs from predators; zooplankters may leave a transient swarm of nitrogen-rich water in their wake; aphids may attract ants to a plant and the ants then prey on other insects there; unsuccessful predators may harass their prey so that they move more and spread a plant virus; or a species may have an unexpected food resource.

In order to detect these linkages in our system it is necessary to know what effects they would have if they are present. Therefore we build a large set of alternative models and identify these responses to changes of parameters which are different in different graphs. These become the objects of observation and experiment.

6 Dynamical Systems and Loop Analysis

In this chapter we present a summary of some mathematics of dynamical systems, descriptions of their behavior, and the correspondence to the loop analysis algorithms. We will be examining primarily linear, continuous, deterministic, nonanticipatory, and time-invariant systems. (See the Appendix for definitions of terms.) In the next chapter we will look at what happens in nonlinear and time-varying systems and introduce time-averaging analysis.

Equilibrium

Any dynamical system, whether linear or nonlinear, can have values for the variables where the rate of change of the variables is zero. These are the equilibrium values.

Given

$$\frac{dy}{dt} = f(y, t; c) \tag{6.1}$$

where y is the dependent variable, t the independent variable, and c is a parameter relating the two variables. Whether or not $f(\cdot)$ is a linear or nonlinear function

$$\frac{dy}{dt} = 0 = f(y, t; c) \tag{6.2}$$

means that for specific values of y and t the rate of change in y with t is zero.

EXAMPLE 6.1

Given

$$\frac{dy}{dt} = ay + b$$

locate the value of y at equilibrium.

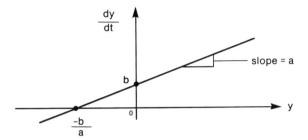

Fig. 6.1

Solution

Plot dy/dt versus y as shown in Figure 6.1. The value of y when $dy/dt = 0$ occurs when $y = -b/a$. Or, mathematically, the equilibrium point can be found as

$$\frac{dy}{dt} = 0 = ay + b$$

$$y^* = \frac{-b}{a}$$

□

EXAMPLE 6.2

Given

$$\frac{dy}{dt} = -ay + by^3$$

locate the value of y at equilibrium.

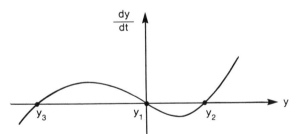

Fig. 6.2

Solution

Plot dy/dt versus y, as in Figure 6.2. Mathematically, the equilibrium points can be obtained as follows.

$$\frac{dy}{dt} = 0 = -ay + by^3$$

Factoring y,
$$y(-a + by^2) = 0$$
Either $y^* = 0$ or $(-a + by^2) = 0$. Therefore
$$y^* = \frac{\pm\sqrt{4ab}}{2b}$$
$$y^* = \pm\sqrt{a/b}$$
The three equilibrium values of y are
$$y_1^* = 0$$
$$y_2^* = +\sqrt{a/b}$$
$$y_3^* = -\sqrt{a/b}$$

The equilbrium points are the real roots of the equation when $dy/dt = 0$. Therefore, the number of equilibrium points cannot exceed the order of the polynomial; or, in a system of equations, the product of their orders. □

Stability

Given the system of Equation 6.3
$$\frac{dy}{dt} = ry - k \qquad (6.3)$$
there is an equilibrium for y at
$$\frac{dy}{dt} = 0 = ry - k$$
$$y^* = \frac{k}{r} \qquad (6.4)$$
The solution to Equation 6.3 is
$$\bar{y} = \bar{y}_0 e^{rt} \qquad (6.5)$$
where
$$\bar{y} = y - y^*, \qquad \bar{y}_0 = y_0 - y^*$$
or
$$y = y_0 e^{rt} + (1 - e^{rt})\left(\frac{k}{r}\right)$$

DYNAMICAL SYSTEMS   153

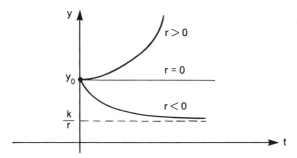

Fig. 6.3

and is shown in Figure 6.3. For $r < 0$ the system moves toward the equilibrium value of $y^* = k/r$. For $r > 0$ the system moves away from the equilibrium value. And, at $r = 0$, the system is static. For this system we can use our intuitive notion of stability. The system is *stable* if it returns to or remains at the equilibrium value. The concept of stability for nonlinear systems will be discussed in the following section. For the present, we will expand on our intuitive ideas of stability with linear systems.

A taxonomy of behavior of a variable with respect to its equilibrium value when plotted over time is illustrated in Figure 6.4.

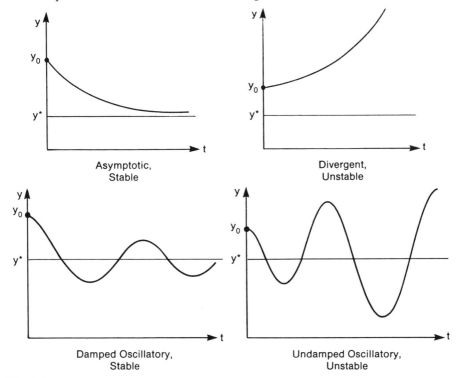

Fig. 6.4

Stability for a two-variable system requires another way of viewing the system. In the system given by

$$y_1 = f_1(y_1, y_2, c_1; t)$$
$$y_2 = f_2(y_1, y_2, c_2; t)$$

each variable has a certain behavior over time. Figure 6.5 illustrates one possibility.

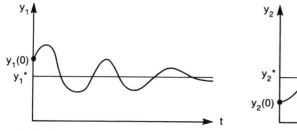
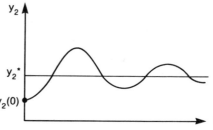

Fig. 6.5

The same two variables also can be plotted on the same graph in relation to each other. Time becomes implicit. This is a phase-plot, as shown in Figure 6.6. This plot shows the *trajectory* of the system. The taxonomy of equili-

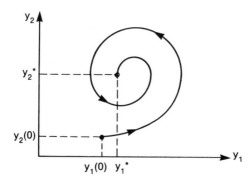

Fig. 6.6

brium points and their associated stability in phase portraits are summarized in Figure 6.7.

For a linear system, stability is a property of the system, that is, the way things are connected. For a nonlinear system, the connections themselves depend on the magnitudes of the variables, so that there can be regions of stable and unstable equilibria. A schematic of a nonlinear system with two regions or domains of stability is shown in Figure 6.8. Note that in a nonlinear system, stability is for a specific region.

DYNAMICAL SYSTEMS 155

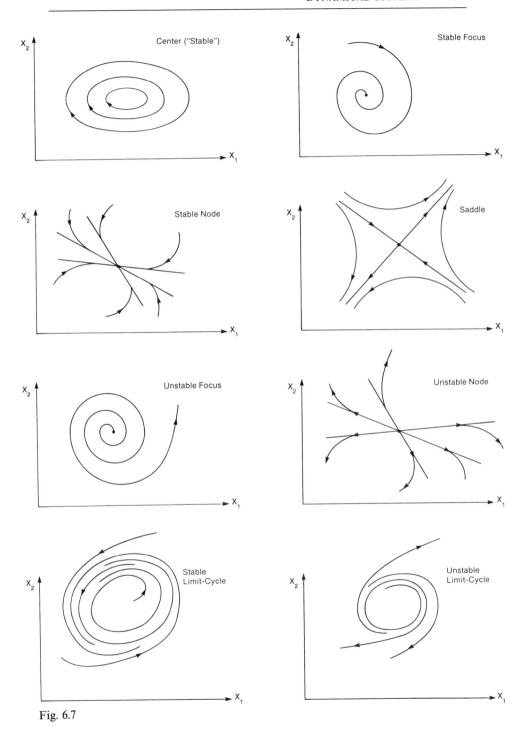

Fig. 6.7

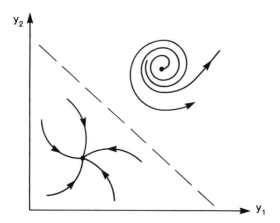

Fig. 6.8

A stable linear system (excluding oscillators) perturbed from its equilibrium point will return to it after some finite amount of time. In contrast, a nonlinear system perturbed from an equilibrium point may return, move to another region in the phase space, exhibit limit cycles, or have one or more of its variables increase unbounded.

The picture of unstable, stable, and semi-stable limit cycles does not correspond with our intuitive notion of stability being only unchanging values of a variable. Variable values are always changing yet in a stable system always returning to within a small value of the equilibrium. For a nonlinear system, stability may refer to the motion of a specific trajectory, not the value at a point.

In what follows, we restrict our analysis to linear systems or to systems that have been linearized around an equilibrium.

In the system of equation 6.6 with several variables, represented by simultaneous equations or higher-order equations, the stability of the system is no longer represented by a single r. The role of r is played by the eigenvalues of the matrix. The solution for y_1 and y_2 in the system

$$\frac{dy_1}{dt} = +r_1 y_1 - k y_2$$
$$\frac{dy_2}{dt} = k y_1 - r_2 y_2$$
(6.6)

or, in matrix notation

$$\begin{bmatrix} \frac{dy_1}{dt} \\ \frac{dy_2}{dt} \end{bmatrix} = \begin{bmatrix} +r_1 & -k \\ k & -r_2 \end{bmatrix} \begin{bmatrix} y_1 \\ y_2 \end{bmatrix}$$
(6.7)

are of the form

$$y_1 = c_{11}e^{\lambda_1 t} + c_{12}e^{\lambda_2 t} \tag{6.8}$$
$$y_2 = c_{21}e^{\lambda_1 t} + c_{22}e^{\lambda_2 t}$$

with the specific values of the c's determined by the initial conditions and the λ's by the values of r_1, r_2, and k.

For a system of n linear differential equations, given in matrix notation by Equation 6.9

$$\dot{\mathbf{X}} = \mathbf{A}\mathbf{X} \tag{6.9}$$

the solution for $\mathbf{X}$ is of the same form as y_1 and y_2 in Equation 6.8. The solution to Equation 6.9 is

$$\mathbf{X} = \mathbf{C}_1 e^{\lambda_1 t} + \mathbf{C}_2 e^{\lambda_2 t} + \mathbf{C}_3 e^{\lambda_3 t} + \cdots \mathbf{C}_n e^{\lambda_n t}$$

where

$$\mathbf{C}_1 = \begin{bmatrix} c_{11} \\ c_{12} \\ \vdots \\ c_{1n} \end{bmatrix}, \quad \mathbf{C}_2 = \begin{bmatrix} c_{21} \\ c_{22} \\ \vdots \\ c_{2n} \end{bmatrix}, \quad \cdots, \quad \mathbf{C}_n = \begin{bmatrix} c_{n1} \\ c_{n2} \\ \vdots \\ c_{nn} \end{bmatrix}$$

The λ's are the eigenvalues of the system, obtained by solving the matrix equation

$$\mathbf{A}\mathbf{X} - \lambda \mathbf{X} = 0$$

or

$$[\mathbf{A} - \lambda \mathbf{I}]\mathbf{X} = 0$$

Since $\mathbf{X} \neq 0$, then

$$|\mathbf{A} - \lambda \mathbf{I}| = 0 = |\lambda \mathbf{I} - \mathbf{A}| \tag{6.10}$$

Equation 6.10 is the characteristic polynomial equation of a linear system. When the determinant is expanded, Equation 6.10 becomes

$$\alpha_0 \lambda^n + \alpha_1 \lambda^{n-1} + \alpha_2 \lambda^{n-2} + \cdots + \alpha_{n-1}\lambda + \alpha_n = 0 \tag{6.11}$$

We can assume $\alpha_0 = 1$ because we can always divide both sides of the polynomial equation 6.11 by α_0. Thus, in slightly simplified form the characteristic equation is

$$\lambda^n + \alpha_1 \lambda^{n-1} + \alpha_2 \lambda^{n-2} + \cdots \alpha_n = 0 \tag{6.12}$$

The general form of the solution for λ will be

$$\lambda = p + iq$$

where $i \equiv \sqrt{-1}$.

For $\lambda = p(q = 0)$, then

$$e^{pt} = \begin{cases} \to \infty, & p > 0 \\ \to 0, & p < 0 \end{cases}$$

For $\lambda = iq(p = 0)$, and from any standard calculus text for Euler's formula relating an exponential with the sine and cosine functions, we can get

$$e^{iqt} = \cos qt + i \sin qt$$

Clearly, any system whose eigenvalues are purely imaginary ($p = 0$) will exhibit oscillatory behavior.

When the eigenvalue is complex, both the real (p) and imaginary (iq) part are present, and $\lambda = p + iq$, then

$$e^{(p+iq)t} = e^{pt}e^{iqt} = e^{pt}(\cos qt + i \sin qt)$$

$$= \begin{cases} \to \infty \text{(growing oscillations)}, & p > 0 \\ \to 0 \text{(diminishing oscillations)}, & p < 0 \end{cases}$$

Hence stability depends on the real part of λ. It is sometimes convenient to use the following notation to denote the real and imaginary parts of the eigenvalues;

$$\text{Re}(\lambda) = p; \quad \text{Im}(\lambda) = q$$

If the system of equations is nonlinear, λ is not a constant of the equation; it is the first partial derivative of the equation with respect to the variable.

EXAMPLE 6.3

Determine the solution of the equation

$$\frac{dX}{dt} = aX(1 - bX) \tag{E6.3.1}$$

and clearly identify the value of λ.

Solution

Write E6.3.1 as

$$\frac{dX}{aX(1 - bX)} = dt \tag{E6.3.2}$$

Then the solution is found by using partial fraction expansion and integrating the terms (see Pielou, 1969, p. 21), which gives

$$X(t) = \frac{X_0 1/b}{1 + e^{-a(t-t_0)}} \tag{E6.3.3}$$

For positive values of a, $X(t)$ approaches the value $1/b$ as t becomes very large; X vanishes with time when a is negative. The equilibrium value of E6.3.1 is either $X^* = 0$ or $X^* = 1/b$. The partial derivative of dX/dt with respect to X evaluated at equilibrium is

$$\frac{\partial}{\partial X}\left(\frac{dX}{dt}\right)_{X=X^*} = (a - 2abX)_{X=X^*}$$

$$= \begin{cases} a & X^* = 0 \\ -a & X^* = 1/b \end{cases}$$

The partial derivative corresponds to the exponent in (6.8), the λ value that determines stability. □

Self-Effect Terms (Self-Damping)

The loop diagram has a one-to-one correspondence with the linear systems matrix. The self-effect loops are the diagonal terms of the matrix. From a straightforward mathematical manipulation we can find how the diagonal terms arise.

Suppose we have a linear, first-order differential equation, as given in Equation 6.13.

$$\begin{aligned} \dot{x}_1 &= -a_1 x_1 - b_1 x_2 - c_1 x_3 \\ \dot{x}_2 &= -a_2 x_1 - b_2 x_2 \\ \dot{x}_3 &= -c_3 x_3 \end{aligned} \tag{6.13}$$

In matrix form this equation becomes

$$\begin{bmatrix} \dot{x}_1 \\ \dot{x}_2 \\ \dot{x}_3 \end{bmatrix} = \begin{bmatrix} -a_1 & -b_1 & -c_1 \\ -a_2 & -b_2 & 0 \\ 0 & 0 & -c_3 \end{bmatrix} \begin{bmatrix} x_1 \\ x_2 \\ x_3 \end{bmatrix} \tag{6.14}$$

For a nonlinear equation, we linearize using the Jacobian (see Appendix), which will give diagonal terms whenever the term

$$\frac{\partial f_i}{\partial x_i}$$

160   DYNAMICAL SYSTEMS

is present, and nonzero when evaluated at the equilibrium. For the general relation,

$$\dot{x}_1 = f_1(x)$$

with $f_1(x^*) = 0$ as an equilibrium, then

$$\frac{\partial \dot{x}_1}{\partial x_1} = \left(\frac{\partial f_1}{\partial x_1}\right)_{x^*}$$

When the qualitative arguments do not reveal if there is self-damping, we formulate the relation of the variable to the rest of the system with a rough or imprecise equation; examine the first partial derivative of the variable with itself, evaluated at equilibrium

$$\frac{\partial\left(\frac{dx_i}{dt}\right)}{\partial x_i}\bigg|_{x^*} \equiv a_{ii}$$

The term a_{ii} if $+$ or $-$ indicates a self-effect is present on variable x_i and if zero, the self-link is absent.

EXAMPLE 6.4

Compare Equations E6.4.1 and E6.4.2 in the determination of self-damping on variable x_1.

$$\frac{dx_1}{dt} = f_1(x) = x_1(ax_1 + bx_2 + cx_3 + \cdots) \quad \text{(E6.4.1)}$$

$$\frac{dx_1}{dt} = f_1(x) = x_1(a + bx_2 + cx_3 + \cdots) \quad \text{(E6.4.2)}$$

Solution

The equilibrium points of E6.4.1 are

$$x_1(ax_1 + bx_2 + cx_3 + \cdots) = 0$$

So, either

$$x_1 = 0 \quad \text{and} \quad (a + bx_2 + cx_3 + \cdots) = 0 \quad \text{(E6.4.3)}$$

Then

$$\frac{\partial f_1}{\partial x_1} = (ax_1 + bx_2 + cx_3 + \cdots)_{x^*} + a(x_1)_{x^*} \quad \text{(E6.4.4)}$$

Only one of the terms on the right-hand side of Equation E6.4.4 will be zero (ignore the trivial case when x's are zero). Thus $(\partial f_1)/(\partial x_1)$ will have some nonzero value.

The equilibrium for E6.4.2 is given by

$$x_1 = 0 \quad \text{and} \quad (a + bx_2 + cx_3 + \cdots) = 0$$

Then

$$\frac{\partial f_1}{\partial x_1} = (a + bx_2 + cx_3 + \cdots)_{x^*} = 0$$

No matter what the value of x_1, there will be no self-damping on it. The difference between Equation E6.4.1 and E6.4.2 is that in E6.4.1 the function includes x_1 to the second power, while E6.4.2 does not. □

EXAMPLE 6.5

Suppose E6.4.2 is changed to include an additive input, I, such as that coming from migration, of variable x_1, as

$$\frac{dx_1}{dt} = f_1(\mathbf{x}) = I + x_1(a + bx_2 + cx_3 + \cdots) \tag{E6.5.1}$$

The equilibrium values are

$$f_1(\mathbf{x}^*) = 0 = I + x_1(a + bx_2 + cx_3 + \cdots)$$

or

$$(a + bx_2 + cx_3 + \cdots) = \frac{I}{x_1}$$

As long as x_1 is not at or near zero, as would be the case for most ecological systems of interest, then

$$\left(\frac{\partial f_1}{\partial x_1}\right)_{x^*} = (a + bx_2 + cx_3 + \cdots)_{x^*} \neq 0$$

An input in the functional growth rate equation of x_1 will cause self-damping on x_1 even though there is no direct density-dependent regulation! Conversely, a negative input, such as the removal of a fixed number of individuals per unit time, will introduce a positive self-effect. □

Signed Digraphs, the Characteristic Equation, and Feedback

The characteristic equation can be obtained from a signed digraph by first recognizing the one-to-one correspondence between the signed digraph and the system matrix.

EXAMPLE 6.6

Draw the signed digraph for the **A** matrix of Equation E6.6.1.

$$\dot{\mathbf{x}} = \mathbf{A}\mathbf{x} = \begin{bmatrix} a_{11} & a_{12} & a_{13} \\ a_{21} & a_{22} & a_{23} \\ a_{31} & a_{32} & a_{33} \end{bmatrix} \begin{bmatrix} x_1 \\ x_2 \\ x_3 \end{bmatrix} \qquad (E6.6.1)$$

Solution
The solution is illustrated in Figure 6.9. □

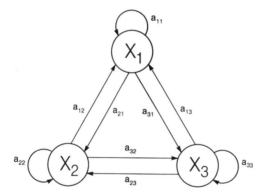

Fig. 6.9

EXAMPLE 6.7

Draw the loop model for the **A** matrix of Equation E6.7.1.

$$\dot{\mathbf{x}} = \mathbf{A}\mathbf{x} = \begin{bmatrix} -a_{11} & +a_{12} & +a_{13} \\ -a_{21} & 0 & -a_{23} \\ 0 & +a_{32} & 0 \end{bmatrix} \begin{bmatrix} x_1 \\ x_2 \\ x_3 \end{bmatrix} \qquad (E.6.7.1)$$

Solution
The solution is illustrated in Figure 6.10. □

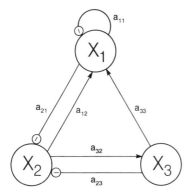

Fig. 6.10

Define $L(m, k)$ as the product of m disjunct loops with k variables (nodes). For the model of Figure 6.10 the possible combinations of m and k and the value of $L(m, k)$ are

$$L(1, 1) = -a_{11}$$

$$L(1, 2) = \begin{cases} -a_{23}a_{32} \\ -a_{21}a_{12} \end{cases}$$

$$L(1, 3) = -a_{21}a_{32}a_{13}$$

$$L(2, 1) \equiv 0$$

$$L(2, 2) = 0$$

$$L(2, 3) = -a_{11}(-a_{23}a_{32})$$

$$L(3, 1) \equiv 0$$

$$L(3, 2) \equiv 0$$

$$L(3, 3) = 0$$

Using the symbol $L(m, k)$, the *determinant* of a signed digraph, and therefore the determinant of the corresponding matrix, can be defined as

$$D_n = \sum_{\text{all } m} (-1)^{n-m} L(m, n)$$

For example, given the determinant of the 2×2 matrix

$$D_2 = \begin{vmatrix} a_{11} & a_{12} \\ a_{21} & a_{22} \end{vmatrix} = a_{11}a_{22} - a_{12}a_{21}$$

The corresponding loop model and determinant are shown in Figure 6.11.

$$D_2 = (-1)^1 \, L(1,2) + (-1)^0 \, L(2,2)$$
$$= -a_{12}a_{21} + a_{11}a_{22}$$

Fig. 6.11

Feedback at Level i is defined to be the coefficient for the term λ^{n-1} in the characteristic polynomial for a matrix. Thus

$$F_k = (-1)^{k+1} D_k$$

The determinant D_k can be expressed in terms of its products of loops, as given in the previous result, to get

$$F_k = \sum_{\substack{\text{for} \\ \text{all } m}} (-1)^{m+1} L(m, k)$$

Therefore if all m loops are negative in a particular term, that term contributes a negative component to F_k since the sign is

$$(-1)^m (-1)^{m+1} = -1.$$

Feedback at Level 0 is defined as

$$F_0 \equiv -1$$

The characteristic polynomial now can be written in terms of the feedback notation

$$F_0 \lambda^n + F_1 \lambda^{n-1} + F_2 \lambda^{n-2} \cdots + F_{n-1} \lambda + F_n = 0 \qquad (6.15)$$

The relation between the coefficients of the characteristic equation and its roots, the values of λ: $\lambda_1, \lambda_2, \lambda_3, \cdots, \lambda_n$, comes from the standard theory of equations. A property of any polynomial, including the characteristic polynomial, is that the coefficients are related to the roots of the equation in a systematic way. Given the characteristic polynomial of Equation (6.11)

$$a_0 \lambda^n + a_1 \lambda^{n-2} + a_2 \lambda^{n-2} + a_3 \lambda^{n-3} + \cdots + a_{n-1} \lambda + a_n = 0$$

the relation between the coefficients a_i and the roots of the equation λ_i can be written as:

$$\frac{a_1}{a_0} = -\sum (\text{all roots})$$

$$\frac{a_2}{a_0} = +\sum (\text{product of the roots taken two at a time})$$

$$\frac{a_3}{a_0} = -\sum (\text{product of the roots taken three at a time})$$

$$\vdots$$

$$\frac{a_n}{a_0} = (-1)^n \sum (\text{product of all } n \text{ roots})$$

Suppose $a_0 = 1$. (If not, divide the characteristic equation by a_0 to get a "new" $a_0 = 1$.) The above property of polynomials becomes

$$a_1 = -\sum_{i=1}^{n} \lambda_i = -(\lambda_1 + \lambda_2 + \lambda_3 + \cdots + \lambda_n)$$

$$a_2 = +\sum_{\substack{i,j=1 \\ i<j}}^{n} \lambda_i \lambda_j$$

$$= \lambda_1\lambda_2 + \lambda_1\lambda_3 + \cdots + \lambda_1\lambda_n + \lambda_2\lambda_3 + \lambda_2\lambda_4 + \cdots$$
$$\lambda_3\lambda_4 + \lambda_3\lambda_5 + \cdots \lambda_{n-1}\lambda_n$$

$$a_3 = -\sum_{\substack{i,j,k=1 \\ i<j<k}}^{n} \lambda_i \lambda_j \lambda_k$$

$$= -(\lambda_1\lambda_2\lambda_3 + \lambda_1\lambda_2\lambda_4 + \lambda_1\lambda_2\lambda_5 + \cdots + \cdots \lambda_1\lambda_3\lambda_4$$
$$+ \lambda_1\lambda_3\lambda_5 + \cdots \lambda_2\lambda_3\lambda_4 + \lambda_2\lambda_3\lambda_5 + \cdots + \cdots + \lambda_{n-2}\lambda_{n-1}\lambda_n)$$

$$\vdots$$

$$a_n = (-1)^n \prod_{i=1}^{n} \lambda_i$$

$$= (-1)^n (\lambda_1 \lambda_2 \lambda_3 \cdots \lambda_n)$$

For stability, the $\text{Re}(\lambda_i) < 0 (i = 1, n)$. A *necessary* condition on the *coefficients* (a's) of the polynomial is that they all have the same sign.

The necessary condition can be seen from the following.

$$a_1 = -\sum_{i=1}^{n} \lambda_i \to \text{Re}(a_1) > 0, \quad \text{since all } \lambda_i < 0.$$

$$a_2 = +\sum_{\substack{i,j=1 \\ i<j}}^{n} \lambda_i \lambda_j \to \text{Re}(a_2) > 0, \quad \text{since all pairwise products of the } \lambda\text{'s} > 0.$$

$$a_3 = -\sum_{\substack{i,j,k=1 \\ i<j<k}}^{n} \lambda_i \lambda_j \lambda_k \to \text{Re}(a_2) > 0, \quad \text{since all triplicate products of the } \lambda\text{'s} < 0.$$

That is, if all the eigenvalues are real, then the product of three negatives is negative and the change of sign makes the overall result positive. If the eigenvalues have complex roots then the polynomial with real coefficients has complex conjugate roots. So, for the complex conjugates $\lambda_1 = -r + is$ and $\lambda_2 = -r - is$ (where $i = \sqrt{-1}$), and the real eigenvalue $\lambda_3 = -r_3$, then

$$-\lambda_1 \lambda_2 \lambda_3 = -(-r + s)(-r - s)(-r_3)$$
$$= -(r^2 + s^2)(-r_3) > 0$$
$$\vdots$$

$$a_n = (-1)^n \prod_{i=1}^{n} \lambda_i \to \text{Re}(a_n) > 0,$$

since for n odd, $\prod \lambda_i < 0$ and $(-1)^n < 0$, therefore $\text{Re}(a_n) > 0$; for n even, $\prod \lambda_i > 0$ and $(-1)^n > 0$, therefore $\text{Re}(a_n) > 0$.

The argument made above remains the same if we take the polynomial equation

$$F(\lambda) = a_0 \lambda^n + a_1 \lambda^{n-1} + \cdots + a_{n-1} \lambda + a_n = 0$$

and multiply it by -1. Thus

$$-F(\lambda) = -a_0 \lambda^n - a_1 \lambda^{n-1} - \cdots - a_{n-1} \lambda - a_n = 0 \qquad (6.16)$$

For the coefficients $(a_i, i = 1, n)$ of Equation 6.16 to have the same sign means they are all negative.

EXAMPLE 6.8

Show that the coefficients for a fourth-order equation

$$(\lambda - \lambda_1)(\lambda - \lambda_2)(\lambda - \lambda_3)(\lambda - \lambda_4) = 0$$

follow the Routh property for the coefficients of polynomials.

Solution

$$(\lambda - \lambda_1)(\lambda - \lambda_2)(\lambda - \lambda_3)(\lambda - \lambda_4) = a_0 \lambda^4 + a_1 \lambda^3 + a_2 \lambda^2 + a_3 \lambda + a_4$$
$$= 0 \qquad (E6.8.1)$$

$$(\lambda^2 - \lambda \lambda_1 - \lambda \lambda_2 + \lambda_1 \lambda_2)(\lambda - \lambda_3)(\lambda - \lambda_4) = 0$$
$$[\lambda^3 - (\lambda_1 + \lambda_2 + \lambda_3)\lambda^2 + (\lambda_1 \lambda_2 + \lambda_1 \lambda_3 + \lambda_2 \lambda_3)\lambda$$
$$- \lambda_1 \lambda_2 \lambda_3](\lambda - \lambda_4) = 0$$
$$\lambda^4 - (\lambda_1 + \lambda_2 + \lambda_3 + \lambda_4)\lambda^3$$
$$+ (\lambda_1 \lambda_2 + \lambda_1 \lambda_3 + \lambda_1 \lambda_4 + \lambda_2 \lambda_3 + \lambda_2 \lambda_4 + \lambda_3 \lambda_4)\lambda^2$$
$$- (\lambda_1 \lambda_2 \lambda_4 + \lambda_1 \lambda_3 \lambda_4 + \lambda_2 \lambda_3 \lambda_4)\lambda + (\lambda_1 \lambda_2 \lambda_3 \lambda_4) = 0 \qquad (E6.8.2)$$

Associate the coefficients in parentheses in E6.8.2 with the corresponding a terms in the first polynomial equation (E6.8.1)

$$1 = a_0 > 0$$
$$-(\lambda_1 + \lambda_2 + \lambda_3 + \lambda_4) = a_1 > 0$$
$$\lambda_1\lambda_2 + \lambda_1\lambda_3 + \lambda_1\lambda_4 + \lambda_2\lambda_4 + \lambda_3\lambda_4 = a_2 > 0$$
$$-(\lambda_1\lambda_2\lambda_3 + \lambda_1\lambda_3\lambda_4 + \lambda_2\lambda_3\lambda_4) = a_3 > 0$$
$$+\lambda_1\lambda_2\lambda_3\lambda_4 = a_4 > 0 \qquad \square$$

The algorithm suggested above can be put more succinctly, as described by the Routh–Hurwitz tabulation

$$
\begin{array}{cccc}
a_0 & a_2 & a_4 & a_6 \cdots \\
a_1 & a_3 & a_5 & a_7 \cdots \\
\dfrac{a_1 a_2 - a_0 a_3}{a_1} = A & \dfrac{a_1 a_4 - a_0 a_5}{a_1} = B & C & D \cdots \\
\dfrac{A a_3 - a_1 B}{A} = E & \dfrac{A a_5 - a_1 C}{A} = F & \cdot & \cdot \\
\dfrac{EB - AF}{E} = G & \cdot & \cdot & \cdot \\
\vdots & \vdots & \vdots & \vdots
\end{array}
$$

The roots of a polynomial either will be *all less* or *all greater* than zero if *all* the elements of the first column are of the same sign; if they are not *all* of the same sign, then the number of positive real values is equal to the number of sign changes.

A somewhat easier formula to remember than the table algorithm is given by the Hurwitz determinants.

The necessary and sufficient condition that all of the real parts of the roots of the characteristic equation (that is, Re (λ_i), for *all* i) are negative is that each of the successive Hurwitz determinants, H_k, $k = 1, n$ all must be positive. (Strictly speaking, they need only be of the same sign, but since we have chosen to represent the characteristic polynomial with a positive first coefficient, this determines the sign criteria.) The Hurwitz determinants are formed as shown

$$H_1 = a_1$$

$$H_2 = \begin{vmatrix} a_1 & a_3 \\ a_0 & a_2 \end{vmatrix}$$

$$H_3 = \begin{vmatrix} a_1 & a_3 & a_5 \\ a_0 & a_2 & a_4 \\ 0 & a_1 & a_3 \end{vmatrix}$$

$$\vdots$$

$$H_n = \begin{vmatrix} a_1 & a_3 & a_5 & a_7 & a_9 & a_{11} & \cdots & a_{2n-1} \\ a_0 & a_2 & a_4 & a_6 & a_8 & a_{10} & \cdots & a_{2n-2} \\ 0 & a_1 & a_3 & a_5 & a_7 & a_9 & \cdots & a_{2n-3} \\ 0 & a_0 & a_2 & a_4 & a_6 & a_8 & \cdots & a_{2n-4} \\ 0 & 0 & a_1 & a_3 & a_5 & a_7 & \cdots & a_{2n-5} \\ 0 & 0 & a_0 & a_2 & a_4 & a_6 & \cdots & a_{2n-6} \\ \cdot & \cdot & \cdot & \cdot & \cdot & \cdot & \cdots & \cdot \\ 0 & 0 & 0 & 0 & 0 & 0 & \cdots & a_n \end{vmatrix}$$

Although the characteristic equation cannot in general be solved for the exact values of λ when the polynomial is fifth-order or higher, the sign of the real part of the eigenvalues can be determined using the Routh–Hurwitz criterion. The sign of the roots of any polynomial, not just those arising from stability matrices of linear differential equations, follow the Routh–Hurwitz criterion. This criterion is as given above.

The determination of stability becomes more understandable with a modified form of the Routh–Hurwitz criterion due to Lienard and Chipart (see Gantmacher, 1960, p. 221). If all the F_k are negative, then only alternate Hurwitz determinants need to be tested. The criterion then becomes (remembering that $a_0 = -1$)

1. $a_k < 0$, for all k.
2. Alternate Hurwitz determinants up to order n are positive. For $n = 3$ we have only to test H_2. For $n = 4$, H_3 is sufficient since $H_1 > 0$ due to condition 1.

In this form, the condition for stability has two parts. The first criterion applies to each a_k separately: If feedback is negative at every level then there cannot be any real positive roots. The second criterion examines the relations among the negative feedbacks at different levels. If it is also satisfied there cannot be any complex roots with positive real parts.

We now relate the criterion to the loop analysis stability criteria using the feedback notation.

$$H_1 = -F_1 > 0$$

$$H_2 = \begin{vmatrix} -F_1 & -F_3 \\ -F_0 & -F_2 \end{vmatrix} > 0$$

$$H_3 = \begin{vmatrix} -F_1 & -F_3 & -F_5 \\ -F_0 & -F_2 & -F_4 \\ 0 & -F_1 & -F_3 \end{vmatrix} > 0$$

$$\vdots$$

$$H_n = \begin{vmatrix} -F_1 & -F_3 & -F_5 & \cdots & -F_{2n-1} \\ -F_0 & -F_2 & -F_4 & \cdots & -F_{2n-2} \\ 0 & -F_1 & -F_3 & \cdots & -F_{2n-3} \\ 0 & -F_0 & -F_2 & \cdots & -F_{2n-4} \\ 0 & 0 & -F_1 & \cdots & -F_{2n-5} \\ 0 & 0 & -F_0 & \cdots & -F_{2n-6} \\ \cdot & \cdot & \cdot & \cdots & \cdot \\ \cdot & \cdot & \cdot & \cdots & \cdot \\ \cdot & \cdot & \cdot & \cdots & \cdot \\ \cdot & \cdot & \cdot & \cdots & -F_n \end{vmatrix} > 0$$

The expansion of the Hurwitz determinants leads to an algebraic form of the second stability criterion in terms of feedbacks or loops; to obtain the loop form, as in Chapter 5, Example 5.5, let L_i be a loop with only variable i (and therefore of length 1), L_{ij} a loop of length 2 with variables i and j, and so on. Then the second criterion for stability for the first two Hurwitz determinants is

$$H_1 = -F_1 > 0$$
$$H_2 = F_1 F_2 + F_3 > 0$$

In terms of loops, H_2 is found as follows.

$$F_1 F_2 = -\sum L_i \sum L_i L_j - \sum L_i \sum L_j L_k + \sum L_i \sum L_{ij} + \sum L_i \sum L_{jk}$$
$$F_3 = \sum L_i L_j L_k - \sum L_i L_{jk} + \sum L_{ijk}$$

But for distinct i, j, k

$$\sum L_i \sum L_j L_k = 3 \sum L_i L_j L_k$$

since the same triplet arises from $(L_i)(L_{jk})$, $(L_j)(L_i L_k)$, and $(L_k)(L_i L_j)$. Therefore

$$H_2 = F_1 F_2 + F_3 = -2 \sum L_i L_j L_k - \sum L_i^2 L_j + \sum L_i L_{ij} + \sum L_{ijk}$$

If all loops are negative, the first three terms are positive and the last one negative. Thus negative loops of length 3 tend to destabilize while the combinations of loops of length 1 and 2 stabilize. But loops of length 2 only appear if conjunct with loops of length 1. In the next example, Example 6.9, the loop of length 2 is disjunct from the only loop of length 1 and therefore does not appear. The first inequality of the second stability criterion is

$$F_1 F_2 + F_3 = L_{123} < 0$$

The next Hurwitz determinant expanded in terms of feedback is

$$H_3 = -F_3 H_2 + F_1 (F_1 F_4 + F_5)$$

but if all the $F_i < 0$ we do not need this condition. The fourth Hurwitz determinant yields

$$H_4 = -F_4 H_3 + H_2 [F_1 F_6 + F_7 - F_5 (F_1 F_4 + F_5)]$$

If the previous criteria are satisfied, the first term is positive. Since H_2 is positive, H_4 will be positive if $F_1 F_6 + F_7$ and $F_1 F_4 + F_5$ are both positive, although that is not necessary. Terms of the form $F_1 F_k + F_{k+1}$ keep recurring in the expansion of the Hurwitz determinant. Consider $F_1 F_4 + F_5$. The product of five distinct loops of length 1 occurs in both terms. But in $F_1 F_4$ it can occur in five ways depending on which loop comes from F_1. The product of four loops in F_4 enters with a minus sign, so that we are left with $-4 \sum L_i L_j L_k L_l L_m$. Products of three loops of length 1 times a loop of length 2 occur three ways in $F_1 F_4$ and once in F_5, giving $2 \sum L_i L_j L_k L_{lm}$, and so on. Loops of length 5 and products of lengths 3 and 2 occur only in F_5. Finally, $F_1 F_4$ contribute to conjunct loops as well. Therefore

$$\begin{aligned}F_1 F_4 + F_5 = &-4 \sum L_i L_j L_k L_l L_m + 2 \sum L_i L_j L_k L_{lm} + \sum L_i L_j L_{klm} \\ &- \sum L_{ij} L_{klm} + \sum L_{ijklm} - \sum L_i^2 L_j L_k L_m + \sum L_i^2 L_j L_{kl} \\ &+ \sum L_i L_{ij} L_k L_l - \sum L_i L_{ij} L_{kl} - \sum L_i L_j L_{ikl} + \sum L_i L_{ijkl}\end{aligned}$$

Therefore in such a term the combinations of short disjunct loops and the conjunct loops contribute positive terms while the longest loops contribute negative terms.

A general and intuitively meaningful expression for H_n in terms of conjunct and disjunct loops has not yet been found.

EXAMPLE 6.9

Compare loop model and matrix analysis for the system depicted in Figure 6.12 and given by Equations E6.9.1 and E6.9.2.

$$\dot{\mathbf{x}} = \mathbf{A}\mathbf{x} \qquad (E6.9.1)$$

$$\mathbf{A} = \begin{bmatrix} 0 & 0 & a_{13} \\ a_{21} & -a_{22} & 0 \\ -a_{31} & -a_{32} & 0 \end{bmatrix} \qquad (E6.9.2)$$

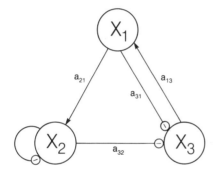

Fig. 6.12

Solution

Loop Analysis

Stability Calculation

Condition 1

$$F_0 \equiv -1 < 0$$
$$F_1 = -a_{32} < 0$$
$$F_2 = -a_{13}a_{31} < 0$$
$$F_3 = -(a_{22})(a_{13}a_{31}) - (a_{21}a_{32}a_{12}) < 0$$

Condition 2

$$F_1F_2 + F_3 > 0$$
$$[a_{22}][a_{13}a_{31}] + [-a_{22}a_{13}a_{31} - a_{21}a_{32}a_{13}] < 0$$

(therefore unstable).

Matrix Analysis

Calculation of the Characteristic Polynomial of E6.9.2

$$\det [\mathbf{A} - \lambda \mathbf{I}] = \begin{vmatrix} -\lambda & 0 & a_{13} \\ a_{21} & -a_{22} - \lambda & 0 \\ -a_{31} & -a_{32} & -\lambda \end{vmatrix} = 0$$

$$-\lambda[(-a_{22} - \lambda)(-\lambda)] + a_{13}[-a_{21}a_{32} - (a_{31})(a_{32} + \lambda)] = 0$$
$$F(\lambda) = \lambda^3 + a_{22}\lambda^2 + a_{13}a_{31}\lambda + a_{13}(a_{21}a_{32} + a_{31}a_{22}) = 0$$

or

$$F(\lambda) = p_0\lambda^3 + p_1\lambda^2 + p_2\lambda + p_3 = 0$$

where

$$p_0 = 1$$
$$p_1 = a_{22}$$
$$p_2 = a_{13}a_{31}$$
$$p_3 = a_{13}(a_{21}a_{32} + a_{31}a_{22})$$

The Routh–Hurwitz criteria for stability are:

1. All the p's must be of the same sign. In this case, all the p's > 0. Hence the first Routh–Hurwitz criterion is met.
2. Form the Hurwitz determinant, which must be greater than zero.

$$H_2 = \begin{vmatrix} p_1 & p_3 \\ p_0 & p_2 \end{vmatrix}$$
$$= \begin{vmatrix} p_{22} & p_{13}(p_{21}p_{32} + p_{31}p_{22}) \\ 1 & p_{13}p_{31} \end{vmatrix}$$

$$p_{22}p_{13}p_{31} - p_{13}p_{21}p_{32} - p_{22}p_{13}p_{31} > 0$$
$$-p_{13}p_{21}p_{32} < 0$$

The second Routh–Hurwitz criterion is not met; therefore, the system is unstable. □

The Routh–Hurwitz criteria for stability follow from the property of the eigenvalues of the characteristic polynomial of linear dynamical systems. In addition to indicating stability, the eigenvalues, λ, also measure the sensitivity of a variable to change. In the equation of a variable with a constant input I and a removal rate r,

$$\frac{dX}{dt} = I - rX \tag{6.17}$$

and the equilibrium level is $X^* = I/r$. The bigger the r value, the less sensitive the equilibrium value is to a change in input. For more than one variable, part of the role of r is taken up by the eigenvalues λ and part by F_n, the feedback of the whole system. The formula for the change in equilibrium value of any variable in a system through a change in parameter (Chapter 3),

$$\frac{\partial}{\partial c}\left(\frac{dX_i}{dt}\right) \approx \frac{\text{path} \times \text{complementary feedback}}{F_n}$$

applied to the first-order system of 6.17 shows: The path is always the identity because we are looking at the effect on the only variable; the complement is the null subsystem, $F_0 \equiv -1$; r divides into the input and plays the role of F_n. The last term of the characteristic polynomial for a multivariable system is F_n

$$\lambda^n + F_1 \lambda^{n-1} + F_2 \lambda^{n-2} + \cdots + F_{n-1} \lambda + F_n = 0$$

and from the relation of coefficients of an equation to its roots.

$$F_n = \lambda_1 \lambda_2 \lambda_3 \cdots \lambda_n$$

The feedback of the overall system is the product of all the λ's. Consequently in a multivariable system the product of the eigenvalues is analogous to the removal rate r in a single variable system, in determining the sensitivity of the equilibrium to parameter change.

Eigenvectors

Another property of dynamical systems is their rate of return to equilibrium after a perturbation from it. In the one-variable case we can ask how long it takes for the system to reach half of its final value. The solution to 6.17 is

$$X(t) = \frac{I}{r}(1 - e^{-r(t-t_0)}) + X_0 e^{-r(t-t_0)} \tag{6.18}$$

where t_0 and X_0 are the time of perturbation and the value of X at that time, respectively. For simplicity we assume the starting time is $t_0 = 0$ and the value that X is perturbed to is $X_0 = 0$. The final value of X is $X(\infty)$ which as expected is the equilibrium value

$$X(\infty) = \frac{I}{r}$$

Therefore the time for X to reach half the equilibrium value is found by solving for t

$$\frac{I}{2r} = \frac{I}{r}(1 - e^{-rt})$$

$$\frac{1}{2} = 1 - e^{-rt}$$

$$e^{-rt} = \frac{1}{2}$$

$$-rt = \ln\frac{1}{2} = -\ln 2$$

$$t = \frac{\ln 2}{r}$$

With two or more variables the rate of return of a system to equilibrium after a perturbation depends on which direction the system is pushed—is it X_1 or X_2 or X_3 or a combination of any two variables or all the variables in equal amounts that are pushed from their equilibrium levels? Different eigenvalues (λ's) are important for each of these cases. For example, perturbing a system only along one of the vector combinations of its variables may make only one of the eigenvalues important in determining the return time. If the perturbation is equal in other directions then a combination of eigenvalues is important.

The solution to a set of first-order differential equations

$$\frac{dX_1}{dt} = a_{11}X_1 + a_{12}X_2 + \cdots + a_{1n}X_n$$

$$\frac{dX_2}{dt} = a_{21}X_1 + a_{22}X_2 + \cdots + a_{2n}X_n$$

$$\vdots$$

$$\frac{dX_n}{dt} = a_{n1}X_1 + a_{n2}X_2 + \cdots + a_{nn}X_n$$

is given by the following, where the b's depend on the initial condition.

$$X_1 = b_{11}e^{\lambda_1 t} + b_{12}e^{\lambda_2 t} + \cdots + b_{1n}e^{\lambda_n t}$$
$$\vdots$$
$$X_n = b_{n1}e^{\lambda_1 t} + b_{n2}e^{\lambda_2 t} + \cdots + b_{nn}e^{\lambda_n t}$$

The λ_i are the eigenvalues and

$$\mathbf{v}_i = \begin{bmatrix} c_{1i}e^{\lambda_i t} \\ c_{2i}e^{\lambda_i t} \\ c_{3i}e^{\lambda_i t} \\ \vdots \\ c_{ni}e^{\lambda_i t} \end{bmatrix}$$

is the eigenvector corresponding to λ_i.

A system of variables represented by a vector of differential equations when displaced from equilibrium follows a trajectory governed by the eigenvector equation

$$\frac{d\mathbf{v}}{dt} = \lambda \mathbf{v} \tag{6.19}$$

The vector, although not the original variable of interest, may be given an intuitive interpretation, in which case it is interesting to know which vector returns quickly and which returns slowly to its equilibrium. Eigenvectors are another set of measures of a system. Sometimes they have intuitive meaning about the structure, as for dominant eigenvectors. An examination of the signed-digraph shows the way the signs in the graph are related to eigenvalues and eigenvectors.

The eigenvector is a linear combination of the variables such that

$$\frac{d}{dt}\begin{bmatrix} v_1 \\ v_2 \\ v_3 \\ \vdots \\ v_n \end{bmatrix} = \begin{bmatrix} \lambda_1 v_1 \\ \lambda_2 v_2 \\ \lambda_3 v_3 \\ \vdots \\ \lambda_n v_n \end{bmatrix} \tag{6.20}$$

where
$v_1 = kc_{11}X_1 + kc_{12}X_2 + kc_{13}X_3 + \cdots + kc_{1n}X_n$
$v_2 = kc_{21}X_1 + kc_{22}X_2 + kc_{23}X_3 + \cdots + kc_{2n}X_n$
$v_3 = kc_{31}X_1 + kc_{32}X_2 + kc_{33}X_3 + \cdots + kc_{3n}X_n$
$$\vdots$$
$v_n = kc_{n1}X_1 + kc_{n2}X_2 + kc_{n3}X_3 + \cdots + kc_{nn}X_n \tag{6.21}$

with k any constant; the c_{ij} are unique coefficients determined by the eigenvalue and the a_{ij}. The solution to (6.20) is a vector of simple exponentials,

$$\begin{bmatrix} v_1 \\ v_2 \\ v_3 \\ \vdots \\ v_n \end{bmatrix} = \begin{bmatrix} v_{01}e^{\lambda_1 t} \\ v_{02}e^{\lambda_2 t} \\ v_{03}e^{\lambda_3 t} \\ \vdots \\ v_{0n}e^{\lambda_n t} \end{bmatrix} \qquad (6.22)$$

and for λ_i negative the v_i return to equilibrium—zero, in this case. A small absolute value of λ_i means a slow return; a large value, a fast return. The average value of the λ_i's is the average of the diagonal terms of the fundamental matrix $\mathbf{A}$ (the self-effect terms in a signed digraph) in the equation

$$\frac{d\mathbf{X}}{dt} = \mathbf{A}\mathbf{X}$$

This can be seen readily by observing that for the characteristic polynomial equation (6.11)

$$\lambda^n + a_1\lambda^{n-1} + a_2\lambda^{n-2} + a_3\lambda^{n-3} + \cdots + a_n = 0$$

$$a_1 = -\sum \lambda_i$$

Expanding the determinant of $|\mathbf{A} - \lambda\mathbf{I}|$ we have

$$a_1 = -\sum a_{ii}$$

So

$$\sum \lambda_i = \sum a_{ii} \quad \text{or}$$

$$\frac{1}{n}\sum \lambda_i = \bar{\lambda} = \frac{1}{n}\sum a_{ii} = \overline{a_{ii}}$$

EXAMPLE 6.10

For a 3×3 matrix show the correspondence between the coefficients of the characteristic polynomial, the eigenvalues and the elements of the matrix.

Solution

Let

$$\mathbf{A} = \begin{bmatrix} a_{11} & a_{12} & a_{13} \\ a_{21} & a_{22} & a_{23} \\ a_{31} & a_{32} & a_{33} \end{bmatrix}$$

Then $|\mathbf{A} - \lambda\mathbf{I}| = 0$.

$$\begin{vmatrix} a_{11} - \lambda & a_{12} & a_{13} \\ a_{21} & a_{22} - \lambda & a_{23} \\ a_{31} & a_{32} & a_{33} - \lambda \end{vmatrix} = 0$$

$$(a_{11} - \lambda)\begin{vmatrix} a_{22} - \lambda & a_{23} \\ a_{32} & a_{33} - \lambda \end{vmatrix} - a_{21}\begin{vmatrix} a_{12} & a_{13} \\ a_{32} & a_{33} - \lambda \end{vmatrix}$$

$$+ a_{31}\begin{vmatrix} a_{12} & a_{13} \\ a_{22} - \lambda & a_{23} \end{vmatrix} = 0$$

$$(a_{11} - \lambda)[(a_{22} - \lambda)(a_{33} - \lambda) - a_{23}a_{32}]$$
$$- a_{21}[a_{12}(a_{33} - \lambda) - a_{13}a_{32}]$$
$$+ a_{31}[a_{12}a_{23} - a_{13}(a_{22} - \lambda)] = 0$$

Multiplying further

$$(a_{11} - \lambda)(a_{22} - \lambda)(a_{33} - \lambda) - a_{23}a_{32}(a_{11} - \lambda) - a_{12}a_{21}(a_{33} - \lambda)$$
$$+ a_{13}a_{21}a_{32} - a_{13}a_{31}(a_{22} - \lambda) + a_{12}a_{31}a_{23} = 0$$

Finally, multiply by -1, collect like terms in λ, and get

$$\lambda^3 - (a_{11} + a_{22} + a_{33})\lambda^2$$
$$+ (a_{11}a_{22} + a_{11}a_{33} + a_{22}a_{33} - a_{23}a_{32} - a_{12}a_{21} - a_{13}a_{31})\lambda$$
$$+ a_{11}a_{23}a_{32} + a_{33}a_{12}a_{21} + a_{22}a_{13}a_{31} - a_{11}a_{22}a_{33}$$
$$- a_{13}a_{21}a_{32} - a_{12}a_{31}a_{23} = 0 \quad \text{(E.6.10.1)}$$

For a third-order polynomial we know there will be three roots, $\lambda_1, \lambda_2, \lambda_3$, which can be factored as

$$(\lambda - \lambda_1)(\lambda - \lambda_2)(\lambda - \lambda_3) = 0 \quad \text{(E6.10.2)}$$

Expand E6.10.2 to get

$$\lambda^3 - (\lambda_1 + \lambda_2 + \lambda_3)\lambda^2 + (\lambda_1\lambda_2 + \lambda_1\lambda_3 + \lambda_2\lambda_3)\lambda - \lambda_1\lambda_2\lambda_3 = 0$$
$$\text{(E6.10.3)}$$

Equating terms of the polynomials of Equations E6.10.1 and E6.10.3 produces

$$(a_{11} + a_{22} + a_{33}) = \lambda_1 + \lambda_2 + \lambda_3 \quad \text{(E6.10.4a)}$$

$$a_{11}a_{22} + a_{11}a_{33} + a_{22}a_{33} - a_{23}a_{32} - a_{12}a_{21} - a_{13}a_{31}$$
$$= \lambda_1\lambda_2 + \lambda_1\lambda_3 + \lambda_2\lambda_3 \quad \text{(E6.10.4b)}$$

$$a_{11}a_{23}a_{32} + a_{33}a_{12}a_{21} + a_{22}a_{13}a_{31} - a_{11}a_{22}a_{33} - a_{13}a_{21}a_{32}$$
$$- a_{12}a_{31}a_{23} = \lambda_1\lambda_2\lambda_3 \quad \text{(E6.10.4c)}$$

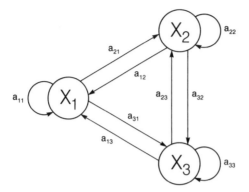

Fig. 6.13

In the last relation (E6.10.4c), the product of the a's is the negative of the determinant of $\mathbf{A}$. The coefficients can also be determined from the feedbacks of the signed-digraph of $\mathbf{A}$, shown in Figure 6.13.

$$-F_1 = (a_{11} + a_{22} + a_{33}) = \lambda_1 + \lambda_2 + \lambda_3$$

$$-F_2 = a_{11}a_{22} + a_{11}a_{33} + a_{22}a_{33} - a_{23}a_{32} - a_{12}a_{21} - a_{13}a_{31}$$
$$= \lambda_1\lambda_2 + \lambda_1\lambda_3 + \lambda_2\lambda_3$$

$$-F_3 = a_{11}a_{23}a_{32} + a_{33}a_{12}a_{21} + a_{22}a_{13}a_{31} - a_{11}a_{22}a_{33}$$
$$- a_{13}a_{21}a_{32} - a_{12}a_{31}a_{23} = \lambda_1\lambda_2\lambda_3 \qquad \square$$

A change in any of the self-effect terms is a change in the average value of the eigenvalues; a change in any of the off-diagonal terms—the links between the variables—has no effect on the average value of the eigenvalues but on a redistribution. As a system changes through increases or decreases in parameter values, new interconnections between variables, or the introduction of new variables, there are changes in one or more of the elements of the system matrix. The resultant redistribution of eigenvalues means some measures become increasingly stable and others less so. In Equation 6.20, with all negative eigenvalues, if $|\lambda_1| > |\lambda_2|$ before any parameter change, so v_1 returns to equilibrium faster than v_2 (Eq. 6.22): The linear combination of the variables which make up the vector v_1 is more stable than the linear combination of v_2. Our problem is to interpret the weighted combination of the variables that make up each vector—the c_{ij}'s.

To find the coefficients c_{ij} of 6.21 is to solve for the cofactors of the characteristic equation matrix

$$\det [\mathbf{A} - \lambda \mathbf{I}] = 0$$

and evaluate the equation at each value of $\lambda_i, j = 1, n$. For distinct values of λ (no two λ's with the same numerical value) there will be n evaluations, each a matrix of n columns proportional to each other. Any one of the columns from each matrix represents the eigenvector for the corresponding eigenvalue. The eigenvectors are the c_{ij}'s for each λ to within a constant (that is, the k value of Equation 6.21 is arbitrary, all of the terms could be multiplied by any same constant number.) Putting the eigenvalues into descending order by magnitude (assuming all negative, real values)

$$|\lambda_1| > |\lambda_2| > |\lambda_3| > \cdots |\lambda_n|$$

then the vector coming from the first matrix produced by the cofactors of $[\mathbf{A} - \lambda \mathbf{I}]$ will reach equilibrium first; then the vector of the second matrix; and so on.

The eigenvector determination can be made from a signed-digraph using a loop analysis algorithm. First, subtract λ from each of the self-terms; when no self-effect loop is present, add a self-damping loop with the coefficient λ. The cofactor of any a_{ij} is the path from variable X_j to X_i multiplied by the complementary feedback. We do not need to worry about division by the overall feedback, as we did in the calculation of change in equilibrium values of variables. Since we are looking for the sign pattern of the vector, division by F_n would divide each term by the same amount; the terms are only in proportion to each other, and arbitrary to within a constant, multiple k of each other.

EXAMPLE 6.11

Find the eigenvalues and associated eigenvectors for the system of a nutrient N, two algal consumers, A_1, A_2, and a specialist herbivore, H.

$$\frac{d}{dt}\begin{bmatrix} N \\ A_1 \\ A_2 \\ H \end{bmatrix} = \begin{bmatrix} -5 & -1 & -2 & 0 \\ 3 & 0 & 0 & 0 \\ 6 & 0 & 0 & -10 \\ 0 & 0 & 2.7 & 0 \end{bmatrix} \begin{bmatrix} N \\ A_1 \\ A_2 \\ H \end{bmatrix}$$

Solution

The eigenvalues are found by solving the characteristic equation

$$|\mathbf{A} - \lambda \mathbf{I}| = 0$$

$$\begin{vmatrix} (-5-\lambda) & -1 & -2 & 0 \\ 3 & -\lambda & 0 & 0 \\ 6 & 0 & -\lambda & -10 \\ 0 & 0 & 2.7 & -\lambda \end{vmatrix} = 0 \quad\quad \text{(E6.11.1)}$$

Expanding around the last column

$$-(-10)\begin{vmatrix}(-5-\lambda) & -1 & -2 \\ 3 & -\lambda & 0 \\ 0 & 0 & 2.7\end{vmatrix} - \lambda\begin{vmatrix}(-5-\lambda) & -1 & -2 \\ 3 & -\lambda & 0 \\ 6 & 0 & -\lambda\end{vmatrix} = 0$$

$$10(2.7)(\lambda^2 + 5\lambda + 3) - \lambda[6(-2\lambda) - \lambda(\lambda^2 + 5\lambda + 3)] = 0$$

$$\lambda^4 + 5\lambda^3 + 42\lambda^2 + 135\lambda + 81 = 0$$

A few guesses for values of λ will quickly reveal $\lambda_1 = -3$ as one solution to the fourth-order equation. We can factor out this term to get the third-order equation

$$(\lambda + 3)(\lambda^3 + 2\lambda^2 + 36\lambda + 27) = 0$$

The third-order equation now can be solved, even if algebraically messy, to get the last three eigenvalues. To two decimal places, all four eigenvalues are

$$\lambda_1 = -3$$
$$\lambda_2 = -0.77$$
$$\lambda_3 = -0.615 + i5.89$$
$$\lambda_4 = -0.615 - i5.89$$

where $i = \sqrt{-1}$.

The four eigenvalues all have negative real values, with the last two complex numbers. The system is stable and the return to equilibrium will follow along an exponential decay for two of the eigenvectors; the two eigenvectors corresponding to the last two eigenvalues will have damped oscillations as they return to equilibrium. The eigenvectors are found from the adjoint of the matrix given in Equation E6.11.1. The adjoint is

Adj $[\mathbf{A} - \lambda \mathbf{I}] =$

$$\begin{bmatrix}
\begin{vmatrix}-\lambda & 0 & 0 \\ 0 & -\lambda & -10 \\ 0 & 2.7 & -\lambda\end{vmatrix} & -\begin{vmatrix}-1 & -2 & 0 \\ 0 & -\lambda & -10 \\ 0 & 2.7 & -\lambda\end{vmatrix} & \begin{vmatrix}-1 & -2 & 0 \\ -\lambda & 0 & 0 \\ 0 & 2.7 & -\lambda\end{vmatrix} & -\begin{vmatrix}-1 & -2 & 0 \\ -\lambda & 0 & 0 \\ 0 & -\lambda & -10\end{vmatrix} \\
-\begin{vmatrix}3 & 0 & 0 \\ 6 & -\lambda & -10 \\ 0 & 2.7 & -\lambda\end{vmatrix} & \begin{vmatrix}(-5-\lambda) & -2 & 0 \\ 6 & -\lambda & -10 \\ 0 & 2.7 & -\lambda\end{vmatrix} & -\begin{vmatrix}(-5-\lambda) & -2 & 0 \\ 3 & 0 & 0 \\ 0 & 2.7 & -\lambda\end{vmatrix} & \begin{vmatrix}(-5-\lambda) & -2 & 0 \\ 3 & 0 & 0 \\ 6 & -\lambda & -10\end{vmatrix} \\
\begin{vmatrix}3 & -\lambda & 0 \\ 6 & 0 & -10 \\ 0 & 0 & -\lambda\end{vmatrix} & -\begin{vmatrix}(-5-\lambda) & -1 & 0 \\ 6 & 0 & -10 \\ 0 & 0 & -\lambda\end{vmatrix} & \begin{vmatrix}(-5-\lambda) & -1 & 0 \\ 3 & -\lambda & 0-\lambda \\ 0 & 0 & 0\end{vmatrix} & -\begin{vmatrix}(-5-\lambda) & -1 & 0 \\ 3 & -\lambda & 0 \\ 6 & 0 & -10\end{vmatrix} \\
-\begin{vmatrix}3 & -\lambda & 0 \\ 6 & 0 & -\lambda \\ 0 & 0 & 2.7\end{vmatrix} & \begin{vmatrix}(-5-\lambda) & -1 & -2 \\ 6 & 0 & -\lambda \\ 0 & 0 & 2.7\end{vmatrix} & -\begin{vmatrix}(-5-\lambda) & -1 & -2 \\ 3 & -\lambda & 0 \\ 0 & 0 & 2.7\end{vmatrix} & \begin{vmatrix}(-5-\lambda) & -1 & -2 \\ 3 & -\lambda & 0 \\ 6 & 0 & -\lambda\end{vmatrix}
\end{bmatrix}$$

Adj $[\mathbf{A} - \lambda \mathbf{I}] =$

$$\begin{bmatrix} (-\lambda^3 - 27\lambda) & (\lambda^2 + 27) & 2\lambda^2 & -20\lambda \\ (-3\lambda^2 - 81) & (-\lambda^3 - 5\lambda^2 - 39\lambda - 135) & 6\lambda & -60 \\ -6\lambda^2 & 6\lambda & (-\lambda^3 - 5\lambda^2 - 3\lambda) & (10\lambda^2 + 50\lambda + 30) \\ -16.2\lambda & 16.2 & (-2.7\lambda^2 - 13.5\lambda - 8.1) & (-\lambda^3 - 5\lambda^2 - 15\lambda) \end{bmatrix}$$

For $\lambda_1 = -3$, the adjoint matrix is

$$\begin{bmatrix} 108 & 36 & 18 & 60 \\ -108 & -36 & -18 & -60 \\ -54 & -18 & -9 & -30 \\ 48.6 & 16.2 & 8.1 & 27 \end{bmatrix} \quad \text{(E6.11.2)}$$

For $\lambda_2 = -0.77$, the adjoint matrix is

$$\begin{bmatrix} 21.2 & 27.6 & 1.2 & 15.4 \\ -82.8 & -107.5 & -4.6 & -60 \\ -3.6 & -4.6 & -0.2 & -2.6 \\ 12.5 & 16.2 & -0.7 & 9.04 \end{bmatrix} \quad \text{(E6.11.3)}$$

For $\lambda_3 = -0.615 + i5.89$, the adjoint matrix is

$$\begin{bmatrix} (-14.17 + i38.62) & (-7.31 - i7.24) & (-68.63 - i14.49) & (12.3 - i117.8) \\ (21.94 + i21.73) & (-3.22 + i4.17) & (-3.69 + i35.34) & -60 \\ (205.88 + i43.47) & (-3.69 + i35.34) & (109.64 + i216.21) & (-343.89 + i222.05) \\ (9.96 - i95.42) & 16.2 & (92.85 - i59.95) & (117.02 + i145.53) \end{bmatrix}$$

$$\text{(E6.11.4)}$$

Lastly, for $\lambda_4 = -0.615 - i5.89$ the adjoint matrix is the same as its complex conjugate λ_3.

The complex adjoint matrix and associated complex eigenvector are easier to manipulate and interpret using the polar or Steinmetz notation for complex numbers. A complex number $x + iy$ can be written as

$r(\cos \theta \pm i \sin \theta)$

where $r = \sqrt{x^2 + y^2}$ and $\theta = \tan^{-1}(y/x)$, with θ usually in degrees. From Euler's formula, an exponential can replace the sine and cosine functions: $e^{\pm i\theta} = \cos \theta \pm i \sin \theta$. The simplified polar representation becomes

$r/\underline{\theta}$

where r is the radius and $/\underline{\theta}$ is a shorthand notation for the exponential $e^{i\theta}$, to indicate the angle θ.

In the polar form the multiplication of two complex numbers is
$$r_1/\underline{\theta_1} \times r_2/\underline{\theta_2} = r_1 r_2/\underline{\theta_1 + \theta_2}$$
For any angle θ the addition or subtraction of $180°$ from it reverses the sign of r. Hence $r/\underline{\theta} = -r/\underline{(\theta - 180°)} = -r/\underline{(\theta - 180°)}$.

(Note: The addition or subtraction of two complex numbers is not the addition of the r's and θ's. To get $r_1/\underline{\theta_1} + r_2/\underline{\theta_2} = r_3/\underline{\theta_3}$ the addition is performed in rectangular coordinates first: $x_3 + i\theta_3 = (x_1 + x_2) + i(\theta_1 + \theta_2)$; then the conversion to polar coordinates is made.)

For the pair of complex conjugate eigenvectors λ_3 and λ_4, the polar representation of the eigenvalues and the complex elements of the adjoint matrix are

$$\lambda_3 = -0.615 + i5.89 = -5.92/\underline{-84}$$
$$\lambda_4 = -0.615 - i5.89 = -5.92/\underline{+84}$$

$$\begin{bmatrix} 61/\underline{141} & 10/\underline{225} & 70/\underline{192} & 118/\underline{-84} \\ 31/\underline{45} & 5/\underline{129} & 36/\underline{96} & 60/\underline{180} \\ 210/\underline{12} & 36/\underline{96} & 242/\underline{63} & 409/\underline{147} \\ 96/\underline{-84} & 16.2/\underline{0} & 111/\underline{-33} & 187/\underline{51} \end{bmatrix}$$

□

EXAMPLE 6.12

Determine the eigenvector sign pattern of the single resource, two algal consumers, and one herbivore model, using the loop analysis algorithm.

Solution

The loop model is shown in Figure 6.14. Subtract λ from each self-effect link (the sign of the subtraction is included in the link symbol, so a self-damping

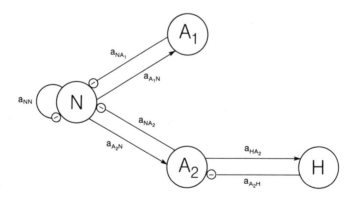

Fig. 6.14

link will incorporate the subtraction; the coefficient of a self-damped link symbol becomes $a_{ii} + \lambda$. The coefficient of a self-enhancing link with λ subtracted from it is $a_{ii} - \lambda$); in the absence of a self-effect link subtract a self-damped link whose coefficient is simply λ. The signed digraph for the eigenvector calculation is shown in Figure 6.15. The negative sign of λ is incorporated into the signed-digraph.

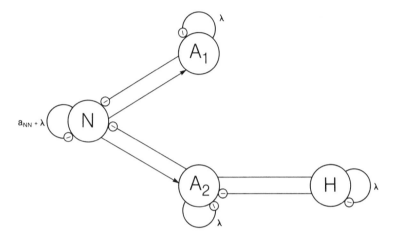

Fig. 6.15

The eigenvector sign pattern algorithm is obtained by using any variable as the starting node. If we use N as the pivotal node, and letting sgn mean the sign pattern but not the magnitude, then

$$\text{sgn}(\mathbf{v}(\lambda)) = \text{sgn} \begin{bmatrix} N^* \\ A_1^* \\ A_2^* \\ H^* \end{bmatrix} = \text{sgn} \begin{bmatrix} p_{NN}[F_3^{(\text{comp})}] \\ p_{A_1N}[F_2^{(\text{comp})}] \\ p_{A_2N}[F_2^{(\text{comp})}] \\ p_{HN}[F_1^{(\text{comp})}] \end{bmatrix}$$

$$\text{sgn}(\mathbf{v}(\lambda)) = \begin{bmatrix} (1)[(\lambda)(-\lambda^2 - a_{A_2H}a_{HA_2})] \\ (a_{A_1N})[(-\lambda^2 - a_{A_2H}a_{HA_2})] \\ (a_{A_2N})[-\lambda^2] \\ (a_{A_2H}a_{HA_2})[-\lambda] \end{bmatrix} = \text{sgn} \begin{bmatrix} v_1(\lambda) \\ v_2(\lambda) \\ v_3(\lambda) \\ v_4(\lambda) \end{bmatrix}$$

Since the model meets all the Routh–Hurwitz stability criteria, all values of λ have negative real parts. Considering only real eigenvalues, the first two elements of the column vector $\mathbf{v}(\lambda)$, that is, $v_1(\lambda)$ and $v_2(\lambda)$, must be of opposite sign from each other; $v_1(\lambda)$ will be positive. The last two elements,

$v_3(\lambda)$ and $v_4(\lambda)$, also will be of opposite sign, with $v_3(\lambda)$ negative. Thus the sign pattern of the eigenvectors associated with real eigenvalues is

$$\mathop{\mathrm{sgn}\,\mathbf{v}(\lambda)}_{\lambda=\lambda_1,\lambda_2,\lambda_3,\lambda_4} = \begin{bmatrix} + \\ - \\ - \\ + \end{bmatrix} \text{ corresponding to } \begin{bmatrix} N \\ A_1 \\ A_2 \\ H \end{bmatrix}$$

This system and that of Example 6.11 are exactly the same, only in this case numbers are not given. The sign pattern is qualitatively the same for both, however (compare this result to E6.11.2). The interpretation is that the ratio of a weighted sum of nutrient N and herbivore H to a weighted sum of the two algae, A_1 and A_2, is the entity coming to equilibrium first in this system, before the variables reach their equilibrium values, which they may never do. □

Each vector returns toward equilibrium at a rate proportional to the real part of its eigenvalue. Therefore the vector which corresponds to the eigenvalue with the largest negative real part returns most rapidly. This will be a measurable property of interest in the system. In Example 6.12, the vector is

$$\mathbf{V} = v_1 N - v_2 A_1 - v_3 A_2 + v_4 H$$

The two algae enter with the same sign and the adjacent levels with opposite sign. Therefore the vector is the difference between the weighted biomasses at alternate levels, the weights depending on their consumption rates and place in the graph structure.

When complex eigenvalues are present, complex-valued eigenvectors emerge. Any eigenvector $\mathbf{v}$ is a solution to the general equation

$$\dot{\mathbf{X}} = \mathbf{A}\mathbf{X} \tag{6.23}$$

so

$$\dot{\mathbf{v}} = \mathbf{A}\mathbf{v} \tag{6.24}$$

A complex eigenvector $v = y + iz$ produces two real solutions to Equation 6.24. Given the eigenvalue $\lambda = a + ib$ for the eigenvector v, the real solutions $\mathbf{X}_1$ and $\mathbf{X}_2$ to 6.24 are

$$\begin{aligned}\mathbf{X}_1(t) &= e^{at}(y \cos bt - z \sin bt) \\ \mathbf{X}_2(t) &= e^{at}(y \sin bt + z \cos bt)\end{aligned} \tag{6.25}$$

The correlation between variables in a complex eigenvector is found from the cosine of the difference in phase angle between them. Two variables that are

90° out of phase will have zero correlation, while two variables 180° out of phase are negatively correlated. From Equation 6.25, we have for the variables of Example 6.11

$$y(t) \approx e^{-0.615t}\left(\begin{bmatrix}-47\\22\\206\\10\end{bmatrix}\cos 5.89t - \begin{bmatrix}39\\22\\43\\-95\end{bmatrix}\sin 5.89t\right),$$

$$z(t) \approx e^{-0.615t}\left(\begin{bmatrix}-47\\22\\206\\10\end{bmatrix}\cos 5.89t + \begin{bmatrix}39\\22\\43\\-95\end{bmatrix}\sin 5.89t\right)$$

All variables continue to oscillate even as they asymptotically approach their equilibrium values, but they become fixed in their relative phase, as can be seen in Figure 6.16.

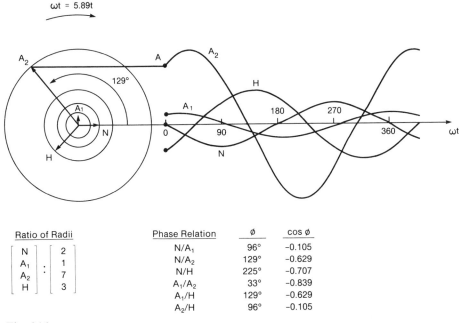

Ratio of Radii		Phase Relation	ø	cos ø
$\begin{bmatrix}N\\A_1\\A_2\\H\end{bmatrix} : \begin{bmatrix}2\\1\\7\\3\end{bmatrix}$		N/A_1	96°	-0.105
		N/A_2	129°	-0.629
		N/H	225°	-0.707
		A_1/A_2	33°	-0.839
		A_1/H	129°	-0.629
		A_2/H	96°	-0.105

Fig. 6.16

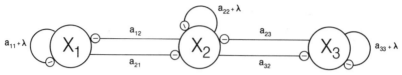

Fig. 6.17

In the community of competitors of Figure 6.17 the sign pattern of the eigenvector is

$$\text{sgn } \mathbf{v}(\lambda) = \text{sgn} \begin{bmatrix} X_1^* \\ X_2^* \\ X_3^* \end{bmatrix} = \text{sgn} \begin{bmatrix} (1)[-(a_{22} + \lambda)(a_{33} + \lambda) - a_{23}a_{32}] \\ (-a_{12})[-(a_{33} + \lambda)] \\ a_{12}a_{32}[-1] \end{bmatrix}$$

For the dominant eigenvalue, λ_{max}, where it is expected that this is a real-valued eigenvalue and $|\lambda_{\text{max}}| > a_{33}$, then

$$\text{sgn } \mathbf{v}(\lambda_{\text{max}}) = \begin{bmatrix} - \\ - \\ - \end{bmatrix}$$

All the terms of the eigenvector are negative. The sign is not important but the fact that they are all the same is, all the signs can be reversed by multiplying each term of the vector by any constant, say -1. This specific sign pattern for the competitors is equivalent to a measure which is some weighted sum of all the variables, and it returns most quickly to equilibrium since it is the quantity associated with the dominant eigenvalue.

The remaining two eigenvalues may be complex conjugates. While in principle the phase relation between competitors and hence the correlation pattern among them can be obtained, without numerical values on the links it is generally not possible to continue without quantification because too many assumptions need to be made.

EIGENVECTORS AND PICTORIAL DESCRIPTIONS

Figure 6.18 is a phase–plane diagram of a hypothetical, two-variable system with a stable equilibrium at (X_1^*, X_2^*). The two lines are the isoclines for the system along which $(dv_1/dt) = 0$, $(dv_2/dt) = 0$.

In the ecological literature the term "isocline" usually refers to the line of no change, as just stated. Any point "above" (or "below") the isocline for the X_2 (ordinate) variable is moving in a "downward" (or "upward") direction

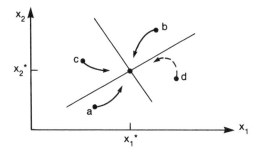

Fig. 6.18

toward the isocline line; a point to the "right" (or "left") of the isocline for the X_1 (abscissa) variable moves in a "leftward" (or "rightward") direction.

The engineering literature uses the term isocline to refer to the set of lines defined by equal slopes, $(dX_2/dX_1) = k$, where k is any constant. It is especially useful in two dimensions for plotting the trajectory of nonlinear systems far from equilibrium. Our use of the term isocline will be recognizable from the context used and will be restricted to either the ecological sense of the line of no change or the mathematical sense of the line parallel to some proportion of the eigenvector line $d(\mathbf{v}/dt) = 0$.

The isoclines generally are not straight lines except in the linear systems case. A perturbation away from the equilibrium point, to a, b, or c in Figure 6.18 will produce trajectories on the return to equilibrium that lie halfway between the two isoclines when the eigenvalues are equal. Point d in Figure 6.18 shows what a trajectory would look like if the eigenvector associated with one of the eigenvalues is stronger than the other. (Of course it is incompatible to have d on the same diagram with a, b, and c other than for pedagogical purposes.)

Two skewed isoclines with unequal eigenvalues ($\lambda_1 < \lambda_2 < 0$) have the trajectories shown in Figure 6.19.

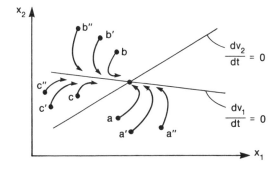

Fig. 6.19

EXAMPLE 6.13

Draw the isoclines and the phase portrait of the linear system given by E6.13.1.

$$\frac{d}{dt}\begin{bmatrix} X_1 \\ X_2 \end{bmatrix} = \begin{bmatrix} -2 & -10 \\ 3 & -15 \end{bmatrix}\begin{bmatrix} X_1 \\ X_2 \end{bmatrix} \tag{E6.13.1}$$

Solution

Find the eigenvalues and eigenvectors.

$$\frac{dX_1}{dt} = 0 = -2X_1 - 10X_2$$

$$\frac{dX_2}{dt} = 0 = 3X_1 - 15X_2$$

$$|\mathbf{A} - \lambda\mathbf{I}| = \begin{vmatrix} (-2-\lambda) & -10 \\ 3 & (-15-\lambda) \end{vmatrix} = 0 \tag{E6.13.2}$$

$$\lambda^2 + 17\lambda + 60 = 0$$

$$\lambda_1 = -12, \quad \lambda_2 = -5$$

The cofactor matrix for $[\mathbf{A} - \lambda\mathbf{I}]$ of E6.13.2 is

$$\begin{bmatrix} -15-\lambda & 10 \\ -3 & -2-\lambda \end{bmatrix}$$

For λ_1, the cofactor matrix is

$$\begin{bmatrix} -3 & 10 \\ -3 & 10 \end{bmatrix}$$

For λ_2, the cofactor matrix is

$$\begin{bmatrix} -10 & 10 \\ -3 & 3 \end{bmatrix}$$

The eigenvectors for λ_1 and λ_2, to within an arbitrary constant, are

$$\mathbf{v}_1 = \begin{bmatrix} 3 \\ 3 \end{bmatrix} \quad \text{and} \quad \mathbf{v}_2 = \begin{bmatrix} 10 \\ 3 \end{bmatrix}$$

(Refer to the graphed solution in Figure 6.20 as well.) □

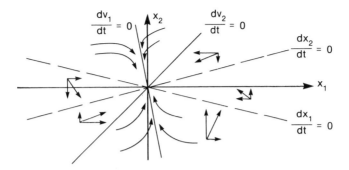

Fig. 6.20

EXAMPLE 6.14

Plot the eigenvector isoclines and the trajectory for the system given by E6.14.1.

$$\frac{d}{dt}\begin{bmatrix} X_1 \\ X_2 \end{bmatrix} = \begin{bmatrix} 4 & 3 \\ 8 & 2 \end{bmatrix}\begin{bmatrix} X_1 \\ X_2 \end{bmatrix} \quad (E6.14.1)$$

Solution

The eigenvalues are $\lambda_1 = -2$, $\lambda_2 = 8$; the system is unstable. The cofactor matrix is

$$\begin{bmatrix} 2 - \lambda & -3 \\ -8 & 4 - \lambda \end{bmatrix}$$

with the two eigenvectors proportional to

$$\mathbf{v}_1 = \begin{bmatrix} 1 \\ -2 \end{bmatrix} \quad \text{and} \quad \mathbf{v}_2 = \begin{bmatrix} 3 \\ 4 \end{bmatrix}$$

The phase plane trajectory for this system is shown in Figure 6.21. Although the system is unstable, nevertheless the variables approach a constant

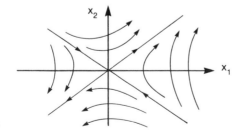

Fig. 6.21

relation with respect to each other; the vector of the dominant eigenvalue λ_2 shows that

$$\begin{bmatrix} X_1 \\ X_2 \end{bmatrix} \text{ is proportional to } \begin{bmatrix} 3 \\ 4 \end{bmatrix}$$

There is a measure of X_1 and X_2 that becomes constant, even though X_1 and X_2 are themselves growing without bound. This constancy for an unstable system will be true only in linear systems. □

Discrete Systems

Life-table analysis divides a population into age-classes, each being the total number of individuals of that age with an age-specific rate of reproduction, mortality, and survival. For instance, Keyfitz (1968) reports that in the United States in 1964 there were 6,546,000 women aged twenty to twenty-four years, with a fertility rate of 0.47607 for female babies per woman for the five-year period and a probability of 0.99605 for a woman in that cohort surviving to the age-class of twenty-five to twenty-nine. Unlike all the previous analyses, which assume continuous rates of change, for example, dX/dt, the mathematical model for life-table analysis frequently assumes discrete processes, $X(t + \Delta t) = f(X(t))$: The value of a variable X at time t plus an increment Δt is a function of the value of X at t, with no change in X until the increment has passed. In the projections for human populations, Δt is typically five or ten years. The well-known Leslie matrix (Keyfitz, 1968; Pollard, 1973), used in population projection, can be extended (Caswell, 1982) to include stage-specific characterizations of populations—organisms that are the same age but in very different states, as a result of complex life-cycles.

A Leslie matrix—or any matrix—can be given a signed-digraph representation, yet both the interpretation of the interactions and stability criteria are altered for discrete-time models. (In some digraphs dynamical systems theory may not even apply; see Roberts, 1976.) In a discrete model, a system is stable if and only if the eigenvalues all lie within a circle of radius one, centered at minus one in the complex plane. We can no longer apply the Routh–Hurwitz criteria for stability. Whether or not a discrete matrix is stable, and in the case of Leslie projection matrices for the United States population it is not, yet there is a set of eigenvectors associated with the eigenvalues and the largest, $\lambda_{\max}$, corresponds to the stable-age distribution.

DYNAMICAL SYSTEMS 191

EXAMPLE 6.15

Determine the stable-age distribution for a hypothetical population with the Leslie matrix

$$\mathbf{A} = \begin{bmatrix} 0 & 5 & 10 \\ 1 & 0 & 0 \\ 0 & 0.4 & 0 \end{bmatrix} \quad \text{(E6.15.1)}$$

Solution

The elements of the Leslie matrix determine the number of offspring produced by members of each age-class and the probability of surviving from one age-class to the next. Clearly **A** is a matrix for a population divided into three age-classes: X_1, X_2, X_3. A population projection one unit time-step, $t + 1$, where a unit of time equals the time-span of the age-classes (for example, an age-class of five years means each unit increment is an increment of five years) is:

$$\begin{bmatrix} X_1(t+1) \\ X_2(t+1) \\ X_3(t+1) \end{bmatrix} = \mathbf{A} \begin{bmatrix} X_1(t) \\ X_2(t) \\ X_3(t) \end{bmatrix} = \begin{bmatrix} 5X_2(t) + 10X_3(t) \\ X_1(t) \\ 0.4X_2(t) \end{bmatrix}$$

It can be seen readily what the elements of the **A** matrix of E6.15.1 mean. The elements of the first row, 5 and 10, are the number of per capita offspring produced that survive by individuals in the second and third age-class, X_1 and X_2, respectively, during one time-interval and who become members of the first generation. The off-diagonal elements 1 and 0.4 indicate the proportion of individuals in the second and third age-class that survive one time-step to become members of the next age-class.

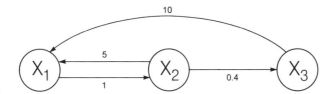

Fig. 6.22

A signed-digraph representation of **A** is depicted in Figure 6.22. The stability of **A** can be determined from a direct solution of the characteristic equation

$$|\mathbf{A} - \lambda \mathbf{I}| = 0$$

$$\begin{vmatrix} -\lambda & 5 & 10 \\ 1 & -\lambda & 0 \\ 0 & 0.4 & -\lambda \end{vmatrix} = 0 \quad \text{or}$$

$$\lambda^3 + -5\lambda + -4 = 0$$

Factoring
$$(\lambda + 1)(\lambda^2 - \lambda - 4) = 0$$
Then
$$\lambda_1 = -1$$
$$\lambda_2 = -1.56$$
$$\lambda_3 = 2.56$$

The eigenvector can be obtained from the λ-modified signed-digraph and loop analysis prescription. Choosing X_1 as the central node, we derive the graph of Figure 6.23

$$\mathbf{v}(\lambda) = \begin{bmatrix} v_1 \\ v_2 \\ v_3 \end{bmatrix} = \begin{bmatrix} -\lambda^2 \\ 5(-\lambda) + 4(-1) \\ 10(-\lambda) \end{bmatrix}$$

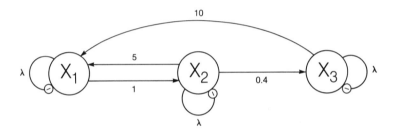

Fig. 6.23

The largest two eigenvalues both exceed ± 1, which violates the stability criteria for linear, discrete systems: The population is growing. Nevertheless, there is a stable age distribution. The age structure will become proportional to:

$$\mathbf{v}(\lambda_3) = \begin{bmatrix} -6.6 \\ -16.8 \\ -25.6 \end{bmatrix}$$

Dividing by the arbitrary constant -6.6 yields

$$\mathbf{v}(\lambda_3) = \begin{bmatrix} 1 \\ 2.6 \\ 3.9 \end{bmatrix}$$

There are roughly two and one-half times as many second-year class individuals and four times as many third-year class individuals as in the first-year class. □

More work with the use of eigenvectors can be developed. The independent identification from a set of data of the vector of minimum variance and its correspondence to the dominant eigenvector has been suggested (Puccia and Levins, 1983). Because the value of the eigenvector analysis is valid mostly around an equilibrium and is restricted to the assumption of linearity to a degree far more than the other topics discussed, we do not expand on it. In the next chapter we take up the question of how to analyze systems far from equilibrium and nonlinear.

7 Time Averaging

Loop analysis treats systems as if they were at a moving equilibrium. That is, the variables are at equilibrium with the parameters that change slowly enough for the variables to keep up with them. In this chapter we consider what happens when we can no longer make that assumption. The variables of interest may have complex trajectories, ones that are periodic or chaotic. If they are bounded, however, they fall within the category of sustained bounded motion (SBM), and what follows applies.

The method of time averaging is used to describe the behavior of the system by looking at average values, measures of variability and of covariance. The measures themselves are defined by analogy to statistical measures which already have fairly clear intuitive meaning. But this does not imply that we are dealing with random processes. Rather, they are simply convenient descriptive measures.

In what follows we first give the definitions and basic rules for time averaging, and introduce the useful manipulations:

    time averaging of a differential equation

    multiplication by a function of X before averaging

    multiplication by dx/dt before averaging

    averaging the derivative of some function of X and Y

    averaging second derivatives

These operations are applied to simple equations with only a few variables. In order to do this, we have to specify the equations much more than we did for loop analysis, although we relaxed the assumption of moving equilibrium.

Initially we assume all processes are continuous (in a mathematical sense) and take place over an "infinite" amount of time. We follow this with a discussion for adjustments when the time interval of observation is finite, and then with a brief consideration of discrete processes.

We end this chapter by examining the relation between loop analysis and time averaging more closely, showing where they give similar results and how the results of loop analysis deal with nonequilibrium equations.

Definitions

The average or expected value of a variable $x(t)$ is $\bar{x}$ or

$$E_t[x] = \frac{1}{t} \int_0^t x(\tau)\, d\tau \tag{7.1}$$

The variance, $\text{Var}_t(x)$, or $\sigma_t^2(x)$ is given by

$$\text{var}_t(x) = \frac{1}{t} \int_0^t [x(\tau) - \bar{x}]^2\, d\tau \tag{7.2}$$

and the covariance between two variables, x and y, is given by

$$\text{cov}_t(x, y) = \frac{1}{t} \int_0^t [x(\tau) - \bar{x}][y(\tau) - \bar{y}]\, d\tau \tag{7.3}$$

The derivative also can be averaged

$$E_t\left[\frac{dx}{dt}\right] = \frac{1}{t} \int_0^t \frac{dx}{d\tau}\, d\tau \tag{7.4}$$

which is

$$E_t\left[\frac{dx}{dt}\right] = \frac{1}{t} [x(t) - x(0)]$$

If the process is bounded, then there are some limits $\max(x)$ and $\min(x)$, and so

$$E_t\left[\frac{dx}{dt}\right] \le \frac{1}{t} [\max(x) - \min(x)] \tag{7.5}$$

Therefore as t increases, $\max(x) - \min(x)$ remains unchanged in the limit

$$\lim_{t \to \infty} E_t\left[\frac{dx}{dt}\right] = E\left[\frac{dx}{dt}\right] = \lim_{t \to \infty} \frac{1}{t} \int_0^t \left[\frac{dx}{dt}\right] = 0 \tag{7.6}$$

That is, the average value of the derivative of a bounded process is zero.

This conclusion holds the following intuitive meaning: If you take a long trip and return to your starting point, the average altitudinal change is zero and the average longitudinal change is zero. If you return to some place near the starting place, these averages are small, and as the duration of the trip increases, the average change approaches zero. For the rest of this section we will take averages in the limit and drop the subscript. Later we will consider averages over finite time intervals.*

---

* The result in 7.6 depends only on the limit of the expression in 7.5 going to zero. If the range $\max(x) - \min(x)$ increases more slowly than t, say $\max(x) - \min(x) \approx \sqrt{t}$, the same result holds.

The expectation operator $E[\cdot]$ can be taken on any differential equation. The averaging operation is used in the same way as common, familiar operators are employed. For example, given an equation

$$x = yz$$

the logarithmic operator $\log(\cdot)$ can be applied to both sides of the equation to obtain

$$\log(x) = \log(yz)$$

which from the definition and properties of logarithms can be simplified to

$$\log(x) = \log(y) + \log(z)$$

to yield a linear relation between the left and right side of the equation.

The expectation operator has its own set of properties that can be used to simplify the relationships among the variables of a differential equation. Thus, with k (any constant), and X, Y variables, the expectation operator properties are:

$$E[k] = k \tag{7.7}$$

$$E[X \pm Y] = E[X] \pm E[Y] = \bar{X} \pm \bar{Y} \tag{7.8}$$

$$E[kX] = kE[X] = k\bar{X} \tag{7.9}$$

$$E[XY] = E[X]E[Y] + \text{cov}(X, Y) = \bar{X}\bar{Y} + \text{cov}(X, Y) \tag{7.10}$$

$$E[X/Y] = E[X]E[1/Y] + \text{cov}(X, 1/Y) = \bar{X}(\overline{1/Y}) + \text{cov}(X, 1/Y) \tag{7.11}$$

$$E[X^2] = E^2[X] + \text{var}(X) = \bar{X}^2 + \text{var}(X) \tag{7.12}$$

Finally, if the average value of a variable is not itself a variable, that is, $\bar{X}$ is a constant, then it follows from 7.7 that

$$E[\bar{X}] = \bar{X} \tag{7.13}$$

Armed with these properties, the averaging operation performed on any differential equation may yield new information about the behavior of the system.

EXAMPLE 7.1

Show that $E[X(K - X)] = \bar{X}(K - \bar{X}) - \text{var}(X)$, where K is a constant and X is a variable.

Solution
First multiply the terms, then average each term separately.
$$E[X(K - X)] = E[KX - X^2]$$
$$= E[KX] - E[X^2]$$

From 7.9 and 7.12 we can obtain
$$E[X(K - X)] = KE[X] - (E^2[X] + \text{var}(X))$$
$$= K\bar{X} - \bar{X}^2 - \text{var}(X)$$
$$= \bar{X}(K - \bar{X}) - \text{var}(X)$$

The same result can be obtained in a more elaborate way, which may help to elucidate the properties of the expectation operator. Let
$$Y = X(K - X)$$

Subtract and add the average value of X and K to all their respective values, which does not change the equation since we are simply adding several zeros to the right-hand side of the equation. (We are treating K as a variable until the end, when we use the fact that it is a constant.)
$$Y = (X - \bar{X} + \bar{X})[(K - \bar{K} + \bar{K}) - (X - \bar{X} + \bar{X})]$$

Collecting terms
$$Y = \{(X - \bar{X}) + \bar{X}\}\{(K - \bar{K}) - (X - \bar{X}) + (\bar{K} - \bar{X})\}$$

Multiply and expand
$$Y = (X - \bar{X})(K - \bar{K}) - (X - \bar{X})(X - \bar{X}) + (X - \bar{X})(K - \bar{X})$$
$$+ \bar{X}(K - \bar{K}) + \bar{X}(X - \bar{X}) + \bar{X}(K - \bar{X})$$

Now average the equation, so
$$E[Y] = E[(X - \bar{X})(K - \bar{K})] - E[(X - \bar{X})(X - \bar{X})]$$
$$+ E[(X - \bar{X})(K - \bar{X})] + E[\bar{X}(K - \bar{K})]$$
$$+ E[\bar{X}(X - \bar{X})] + E[\bar{X}(K - \bar{X})] \quad \text{(E7.1.1)}$$

From Equation 7.10, for any two variables W and Z,
$$\text{cov}(W, Z) = E[WZ] - E[W]E[Z]$$
$$= E[WZ] - \bar{W}\bar{Z} \quad \text{(E7.1.2)}$$

And, from Equation 7.12
$$\text{var}(X) = E[X^2] - \bar{X}^2 \quad \text{(E7.1.3)}$$

Each term of E7.1.1 can be multiplied and evaluated using Equations 7.7, 7.8, 7.13, E7.1.2 and E7.1.3 as follows

$$E[(X - \bar{X})(K - \bar{K})] = E[XK - \bar{X}K - X\bar{K} + \bar{X}\bar{K}] \quad \text{(E7.1.4a)}$$
$$= E[XK] - E[\bar{X}K] - E[X\bar{K}] + E[\bar{X}\bar{K}]$$
$$= E[XK] - \bar{X}\bar{K} - \bar{X}\bar{K} + \bar{X}\bar{K}$$
$$= E[XK] - \bar{X}\bar{K} \quad \text{(E7.1.4b)}$$
$$E[(X - \bar{X})(K - \bar{K})] = \text{cov}(X, K) \quad \text{(E7.1.4c)}$$

Note from E7.1.4c, which is the definition of covariance, that $\text{cov}(X, K) = \text{cov}(K, X)$ since the order of multiplication of the two terms does not alter the results.

We now use the fact that in this case K is a constant, so

$$E[X, K] = KE[X] = K\bar{X} = \bar{K}\bar{X}$$

and therefore

$$\text{cov}(X, K) = 0 \quad \text{(E7.1.5)}$$
$$E[(X - \bar{X})(X - \bar{X})] = E[X^2 - 2\bar{X}X + \bar{X}^2] \quad \text{(E7.1.6)}$$
$$= E[X^2] - 2\bar{X}^2 + \bar{X}^2$$
$$= E[X^2] - \bar{X}^2$$
$$E[(X - \bar{X})(X - \bar{X})] = \text{var}(X) \quad \text{(E7.1.7)}$$
$$E[\bar{X}(K - \bar{X})] = E[\bar{X}K - \bar{X}^2] \quad \text{(E7.1.8)}$$
$$= E[\bar{X}K] - E[\bar{X}^2]$$
$$= K\bar{X} - \bar{X}^2$$
$$E[(X - \bar{X})(K - \bar{X})] = E[XK - \bar{X}K - X\bar{X} + \bar{X}^2] \quad \text{(E7.1.9)}$$
$$= K\bar{X} - K\bar{X} - \bar{X}^2 + \bar{X}^2$$
$$= 0$$
$$E[\bar{X}(K - \bar{K})] = E[\bar{X}K - \bar{X}\bar{K}] \quad \text{(E7.1.10)}$$
$$= \bar{X}\bar{K} - \bar{X}\bar{K}$$
$$= 0$$
$$E[X - \bar{X}] = E[X] - \bar{X} \quad \text{(E7.1.11)}$$
$$= 0$$

Substituting E7.1.5 through E7.1.11 into E7.1.1 gives the result

$$E[Y] = E[X(K - X)] = K\bar{X} - \bar{X}^2 - \text{var}(X)$$
$$= \bar{X}(K - \bar{X}) - \text{var}(X)$$

which is what we set out to prove.

If K had not been a constant but some function of time, $K = K(t)$, then Equation E7.1.5 would not be equal to zero and then

$$E[X(K(t) - X)] = \bar{X}(\bar{K} - \bar{X}) + \text{cov}(X, K) - \text{var}(X)$$

It would be worth keeping Equations E7.1.4c and E7.1.7 handy since these are basic relations of the expectation operator analogous to the definitions of variance and covariance in standard statistics, and will be used frequently.

□

Operations with Time Averaging

Differential equations can be averaged directly or after prior manipulation. Consider first the population growth equation of Example 7.1.

$$\frac{dX}{dt} = X(K - X) \tag{7.14}$$

where K is some constant parameter. The expected value is

$$E\left[\frac{dX}{dt}\right] = \bar{X}(K - \bar{X}) - \text{var}(X) \tag{7.15}$$

which is zero. But we can divide 7.14 by X if X is never too close to zero to get

$$\frac{1}{X}\frac{dX}{dt} = K - X \tag{7.16}$$

Since the left-hand side is the derivative of the natural logarithm $(d(\ln X)/dt) = 1/X \, (dX/dt)$, it also has an expected value of zero. Therefore

$$\bar{X} = K \tag{7.17}$$

and

$$\text{var}(X) = 0 \tag{7.18}$$

That is, X does not vary: There is no sustained bounded motion. This, of course, is a known result. The equation leads to an equilibrium at $X^* = K$ or $X^* = 0$. The demonstration that a variance is equal to zero is one way of excluding sustained bounded motion (SBM).

Now let K be some variable $K(t)$ which may depend on other variables in the system or on external factors. The only change in Equation 7.17 is that $\bar{X} = \bar{K}$, the average value of X is the average value of K. But 7.15 becomes

$$E\left[\frac{dX}{dt}\right] = \bar{X}(\bar{K} - \bar{X}) + \text{cov}(X, K) - \text{var}(X) \tag{7.19}$$

Substituting 7.17 into 7.19 and setting it equal to zero, we have

$$\text{cov}(K, X) = \text{var}(X) \tag{7.20}$$

Therefore the covariance of K with X is positive. Further, we can show that the variance of X is less than the variance of K: From the definition of variance and covariance

$$\sigma_K \sigma_X \rho = \sigma_X^2 \tag{7.21}$$

where ρ is the correlation coefficient and therefore lies between -1 and $+1$. Dividing by σ_X and then squaring the result we get

$$\sigma_K^2 \rho^2 = \sigma_X^2 \tag{7.22}$$

But since ρ^2 is less than or equal to 1, $\sigma_K^2 \geq \sigma_X^2$. This will hold no matter what the dynamics of K may be.

EXAMPLE 7.2

Examine the process

$$\frac{dX}{dt} = XS(t)$$

Find whether SBM is possible, and find the average value of $S(t)$ and the covariance of X with $S(t)$.

Solution

Divide the equation by X

$$\frac{1}{X}\frac{dX}{dt} = S(t)$$

Therefore if X is bounded away from zero and is bounded,

$$E\left[\frac{1}{X}\frac{dX}{dt}\right] = 0 = E[S(t)]$$

But since $S(t)$ can be any function of time, in general it will not have a zero average. Therefore either X is unbounded (if $E[S(t)]$ is positive) or X gets closer and closer to zero so that the division by X is not valid. In either case the equation cannot describe a process with sustained bounded motion. In the special case where $E[S(t)] = 0$, SBM is possible. Then, since

$$E\left[\frac{dX}{dt}\right] = \bar{X}\bar{S} + \text{cov}(X, S)$$

$$\text{cov}(X, S) = 0 \qquad \square$$

In what follows we restrict our analysis to linear systems or to systems that have been linearized around an equilibrium.

EXAMPLE 7.3

For a long time ecologists debated whether population growth was density dependent. It was argued that if population growth declined when the population was large, then there should be a negative correlation between population size and growth rate. Show that this inference does not hold.

Solution
Let

$$\frac{dX}{dt} = R(X, Y) \tag{E7.3.1}$$

be any kind of population growth. If there is sustained bounded motion (SBM) then $\bar{R} = 0$. Multiply Equation E7.3.1 by X. The left-hand side is now

$$X\frac{dX}{dt}$$

which is

$$\frac{1}{2}\left(\frac{dX^2}{dt}\right)$$

Therefore its expected value is zero, so that

$$E\{XR(X, Y)\} = 0$$

The covariance of population size and growth rate is zero! This result seems paradoxical. A closer examination shows, however, that the result depends on SBM: If a population is near maximum density, it has either been increasing recently to reach maximum or has begun to decrease from maximum. High densities occur in association both with increase and decrease. If the population were not in SBM, it might simply increase toward a maximum at a diminishing rate. Then the previous intuition of ecologists would be valid. When we consider expectations over finite time periods, this difference will become clearer. □

In both examples we multiplied the equation by some function of X before taking average values. This can be done with any bounded function of X, say $g(x)$, since $g(X)(dX/dt)$ is the derivative of some function of X. We cannot multiply by a function of any other variable, however. It is not generally true that

$$E\left\{Y\frac{dX}{dt}\right\} = 0$$

EXAMPLE 7.4

Consider a biochemical process in which

$$\frac{dX}{dt} = S - \frac{v_{max}X}{K+X} = 0 \qquad (E7.4.1)$$

where S is the rate of synthesis of a molecule and v_{max} and K are the kinetic constants of a monomolecular enzymatic process that transforms the molecule into something else. Find the covariance of S and X.

Solution

First take expected values and set equal to zero. Thus

$$\bar{S} = v_{max} E\left[\frac{X}{K+X}\right]$$

Now multiply E7.4.1 by $K + X$, then take the average to get

$$(\bar{X} + K)\bar{S} + \text{cov}(S, X) = v_{max}\bar{X} \quad \text{or}$$

$$\bar{S} + \frac{\text{cov}(S, X)}{K+X} = \frac{v_{max}\bar{X}}{K+X}$$

The shape of $(v_{max}\bar{X})/(K+X)$ is convex upward, as shown in Figure 7.1.

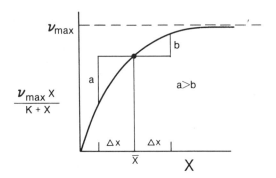

Fig. 7.1

Therefore, any variation about an average $\bar{X}$ increases the function when $X > \bar{X}$ by a smaller amount than it decreases the function when $X < \bar{X}$. Therefore

$$E\left[\frac{X}{K+X}\right] < \frac{\bar{X}}{K+\bar{X}}$$

So

$$\bar{S} + \frac{\text{cov}(S, X)}{K + \bar{X}} > \bar{S} \quad \text{and} \quad \text{cov}(S, X) > 0$$

Similarly, if S were constant but $v_{\max}$ variable, we could show a negative correlation between $v_{\max}$ and X. □

A second kind of manipulation starts with a function of X and Y. Since

$$\frac{d}{dt}(XY) = X\frac{dY}{dt} + Y\frac{dX}{dt} \qquad (7.23)$$

$$E\left[X\frac{dY}{dt}\right] = -E\left[Y\frac{dX}{dt}\right] \quad \text{or} \qquad (7.24)$$

$$\text{cov}\left(X, \frac{dY}{dt}\right) = -\text{cov}\left(\frac{dX}{dt}, Y\right) \qquad (7.25)$$

This has an immediate application to field ecology. Suppose we suspect that X and Y are a predator–prey pair. Then when the prey are abundant, the predators should increase, and when the predators are abundant, the prey should decrease. Therefore the biological argument suggests that $\text{cov}(X, (dY)/(dt))$ should have the opposite sign from $\text{cov}((dX/dt), Y)$. But if we observe it to be true, this is not a confirmation of the supposed biological relation at all; it holds for any pair of variables, interacting or not, that have sustained bounded motion! This is another example of ways in which the commonsense use of statistics would be misleading.

We get further information by applying these techniques to systems of equations. For instance, consider a single resource R and consumer X linked by the equation

$$\frac{dR}{dt} = I - R(pX + c) \qquad (7.26)$$

$$\frac{dX}{dt} = X(pR - \theta) \qquad (7.27)$$

This is the R, X, S system of Chapter 4, in which the physiological state S changes so rapidly that it is in equilibrium with the resource R and need not be identified separately. This system would reach an equilibrium at

$$R^* = \frac{\theta}{p}$$

$$X^* = \frac{I}{\theta} - \frac{c}{p}$$

But if any of the parameters varies, the system can show SBM. Suppose first that $I = I(t)$, a variable input to the community. Dividing Equation 7.27 by X (using the property of Equation 7.9), we find the average value of R by

$$E\left[\frac{1}{X}\frac{dX}{dt}\right] = 0 = E[pR - \theta] = p\bar{R} - \theta \tag{7.28}$$

Or

$$\bar{R} = \frac{\theta}{p} \tag{7.29}$$

From Equation 7.27 and the property of the expectation operator of Equation 7.10

$$E\left[\frac{dX}{dt}\right] = 0 = \bar{X}(p\bar{R} - \theta) + p \text{ cov }(R, X) \tag{7.30}$$

Substituting Equation 7.28 into 7.30 yields

$$\text{cov }(R, X) = 0$$

Applying the expectation operator to Equation 7.26

$$0 = \bar{I} - \bar{R}(p\bar{X} + c) - p \text{ cov }(R, X) \tag{7.31}$$

The covariance is zero so

$$\frac{\bar{I}}{\bar{R}} = p\bar{X} + c \tag{7.32a}$$

Substituting in the value of $\bar{R}$ from 7.28 and rearranging terms yields

$$\bar{X} = \frac{\bar{I}}{\theta} - \frac{c}{p} \tag{7.32b}$$

Finally, dividing Equation 7.26 by R before averaging

$$E\left[\frac{1}{R}\frac{dR}{dt}\right] = 0 = E\left[\frac{I}{R}\right] - E[pX + c]$$

From the property of Equation 7.11

$$0 = I\overline{\left(\frac{1}{R}\right)} + \text{cov}(I, 1/R) - (p\bar{X} + c) \quad \text{and}$$

$$I\overline{\left(\frac{1}{R}\right)} = p\bar{X} + c - \text{cov}(I, 1/R) \tag{7.33a}$$

From 7.32a

$$\frac{\bar{I}}{\bar{R}} = p\bar{X} + c \tag{7.33b}$$

From the general rule that for real values that are not all the same a harmonic mean is greater than the inverse mean; then, if R varies at all,

$$\overline{\left(\frac{1}{R}\right)} > \left(\frac{1}{\bar{R}}\right)$$

Equation 7.33a can be greater than 7.33b only if cov $(I, 1/R) < 0$. Therefore R is correlated positively (since $1/R$ is correlated negatively) with its input but not with its consumer.

Suppose instead that the death rate of X varies, $\theta = \theta(t)$, while I is constant. Now, from Equation 7.27

$$p \text{ cov}(X, R) = \text{cov}(X, \theta) \tag{7.34}$$

From Equation 7.31 (with I now constant)

$$I = \bar{R}(p\bar{X} + c) + p \text{ cov}(R, X) \quad \text{or} \tag{7.35}$$

$$\frac{I}{\bar{R}} = p\bar{X} + c + \frac{p \text{ cov}(R, X)}{\bar{R}}$$

Dividing 7.26 by R and then averaging

$$E\left[\frac{1}{R}\frac{dR}{dt}\right] = 0 = E\left[\frac{I}{R}\right] - E[pX + c]$$

This time I is a constant so

$$I\overline{\left(\frac{1}{R}\right)} = p\bar{X} + c \tag{7.36}$$

But if R varies at all, and from the previously noted harmonic mean relation to the inverse mean

$$\overline{\left(\frac{1}{R}\right)} > \left(\frac{1}{\bar{R}}\right)$$

then Equation 7.35 is less than 7.36, that is

$$p\bar{X} + c + \frac{p \operatorname{cov}(R, X)}{\bar{R}} < p\bar{X} + c \quad \text{therefore}$$

$$\frac{p \operatorname{cov}(R, X)}{\bar{R}} < 0, \quad \text{so that} \quad \operatorname{cov}(R, X) < 0 \tag{7.37}$$

Substituting 7.37 into 7.34

$$\operatorname{cov}(X, \theta) < 0$$

Notice that for the same functional relationship between R and X we can get different covariance patterns depending on where the environmental variation enters the system. In the first case the variation entered through I, the input of resource to the community, and in the second case through the death rate of the consumer.

Consider next a community with two prey and two predators, one of which utilizes both prey, while the other is a specialist. This community is depicted in Figure 7.2.

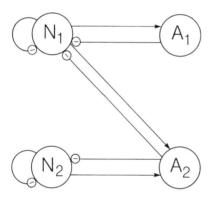

Fig. 7.2

Let the equations be

$$\frac{dN_1}{dt} = N_1(K_1 - p_{11}A_1 - p_{12}A_2 - N_1) \tag{7.38a}$$

$$\frac{dN_2}{dt} = N_2(K_2 - p_{22}A_2 - N_2) \tag{7.38b}$$

$$\frac{dA_1}{dt} = A_1(p_{11}N_1 - \theta_1) \tag{7.38c}$$

$$\frac{dA_2}{dt} = A_2(p_{12}N_1 + p_{22}N_2 - \theta_2) \tag{7.38d}$$

Suppose that variation in the system enters through the variable parameter K_1. Then from the expectation operation on Equation 7.38b,

$$E\left[\frac{dN_2}{dt}\right] = 0$$
$$= \bar{N}_2(K_2 - p_{22}\bar{A}_2 - \bar{N}_2) - p_{22}\,\text{cov}(N_2, A_2) - \text{var}(N_2)$$
(7.39)

Now divide Equation 7.38b by N_2 first before averaging, so

$$E\left[\frac{1}{N_2}\frac{dN_2}{dt}\right] = 0 = (K_2 - p_{22}\bar{A}_2 - \bar{N}_2)$$

Substituting into the first term of Equation 7.39, we find

$$\text{cov}(A_2, N_2) < 0$$

Following a similar procedure of taking the average both on the equation before and after division by the variable of the derivative for Equations 7.38a, b, and d produces the following results.

From Equation 7.38c

$$\text{cov}(A_1, N_1) = 0 \tag{7.40}$$

From Equation 7.38d and the previous results

$$p_{12}\,\text{cov}(A_2, N_1) + p_{22}\,\text{cov}(A_2, N_2) = 0 \tag{7.41}$$

Therefore

$$\text{cov}(A_2, N_1) > 0 \tag{7.42}$$

Finally, from Equation 7.38a

$$\text{cov}(N_1, K_1) - p_{11}\,\text{cov}(A_1, N_1) - p_{12}\,\text{cov}(A_2, N_1) - \text{var}(N_1) = 0$$

Substituting the results of 7.40, and from 7.42 along with the fact that the variance is always positive

$$\text{cov}(N_1, K_1) = p_{12}\,\text{cov}(A_2, N_1) + \text{var}(N_1) > 0 \tag{7.43}$$

Thus A_1 is uncorrelated with its prey, whereas A_2 has a positive correlation with the shared prey and a negative correlation with its exclusive prey. Prey N_1 is positively correlated with its own resource K_1.

These correlations have the following biological implications. Since N_1 is positively correlated both with its own resource and one of its predators, population peaks of N_1 will be periods of abundant resources, high reproduction, and high mortality, resulting in a young population. When N_1 is at a minimum, both reproduction and mortality are low and the population will be older. For N_2 this is reversed. The prey species have very different demographies because of predator specialization.

EXAMPLE 7.5

Determine the correlation pattern of the two prey, two predator community given in Equations 7.38 and shown in Figure 7.1, where the variable parameter is $K_2 = K_2(t)$, the resource of prey N_2.

Solution

The average of 7.38a yields

$$E\left[\frac{dN_1}{dt}\right] = 0 = \bar{N}_1(K_1 - p_{11}\bar{A}_1 - p_{12}\bar{A}_2 - \bar{N}_1)$$
$$- p_{11} \operatorname{cov}(A_1, N_1) - p_{12} \operatorname{cov}(A_2, N_1) - \operatorname{var}(N_1)$$

The average of Equation 7.38a first divided by N_1 yields

$$E\left[\frac{1}{N_1}\frac{dN_1}{dt}\right] = 0 = K_1 - p_{11}\bar{A}_1 - p_{12}\bar{A}_2,$$

or

$$K_1 = p_{11}\bar{A}_1 + p_{12}\bar{A}_2$$

which substituted into the previous result produces

$$-p_{11} \operatorname{cov}(A_1, N_1) - p_{12} \operatorname{cov}(A_2, N_1) = \operatorname{var}(N_1)$$

From the relation of 7.40 and the fact that the variance is always positive

$$-\operatorname{cov}(A_2, N_1) > 0 \quad \text{or} \quad \operatorname{cov}(A_2, N_1) < 0 \quad (\text{E7.5.1})$$

In similar fashion, we obtain from 7.38d

$$p_{22} \operatorname{cov}(A_2, N_2) = -p_{12} \operatorname{cov}(A_2, N_1)$$

so, from the result of 7.43

$$-\operatorname{cov}(A_2, N_2) < 0 \quad \text{or} \quad \operatorname{cov}(A_2, N_2) > 0 \quad (\text{E7.5.2})$$

From Equation 7.38b

$$0 = \operatorname{cov}(N_2, K_2) - p_{22} \operatorname{cov}(A_2, N_2) - \operatorname{var}(N_2) \quad \text{or}$$
$$\operatorname{cov}(N_2, K_2) = p_{22} \operatorname{cov}(A_2, N_2) + \operatorname{var}(N_2)$$

Substituting the results of 7.44 into the above yields

$$\operatorname{cov}(N_2, K_2) > 0 \qquad (E7.5.3)$$

The predator A_1 remains uncorrelated with its shared prey, while A_2 is now negatively correlated with N_1 and positively correlated with its exclusive prey N_2. Hence the correlation pattern of A_2 with the prey items is reversed when variation enters the resource, K_2, of its exclusive prey, from when the variation is via the resource, K_1, of the shared prey. The prey N_2 is positively correlated with its own resource, K_2. □

EXAMPLE 7.6

What is the covariation pattern for the above two prey, two predator system when variation enters through the death rate θ_1 of the specialist predator A_1?

Solution

Averaging 7.38b and 7.38c both before and after dividing by N_1 and A_1, respectively, gives the immediate result

$$\operatorname{cov}(A_2, N_2) = -\operatorname{var}(N_2) < 0, \quad \text{and} \qquad (E7.6.1)$$

$$\operatorname{cov}(A_1, N_1) = 0 \qquad (E7.6.2)$$

Now average 7.38a before and after division and substitute the result in (E7.6.2). This gives

$$\operatorname{cov}(A_2, N_1) = -\operatorname{var}(N_1) < 0$$

Finally, from 7.38d

$$\operatorname{cov}(A_2, N_1) = -\operatorname{var}(N_1) < 0$$

A_1 is still uncorrelated with its prey N_1, as in the previous example, while A_2 is now correlated negatively with both its prey. □

A biological interpretation of the dynamic for the two prey, two predator system described by Equations 7.38a–d is as follows. From Equations 7.43 and E7.5.3 we see that either prey species would be positively correlated with its own resources if these vary. The negative covariance between A_2 and N_2 implies that the product $p_{22} A_2 N_2$, which measures predation, is less than it would be under constant conditions. From Equation 7.39, for example, the covariance $-p_{22} \operatorname{cov}(A_2, N_2)$ enhances the predation term (that is, negative

p_{22} times negative covariance is a positive quantity). If there were no variation in N_2, then

$$E\left[\frac{dN_2}{dt}\right] = 0 = N_2(K_2 - p_{22}A_2 - N_2)$$

which, compared to 7.39, would require the term in parentheses to be smaller, or the predation term $p_{22}A_2N_2$ to be of greater magnitude. N_2 experiences relatively less predation stress and more density regulation by its own numbers because when N_2 is rare, A_2 is abundant, and predation is relatively intense; but when N_2 is abundant, A_2 is rare.

EXAMPLE 7.7

Consider the R, S, X system discussed in Chapter 4, Example 4.6, in which resource R is consumed by a species with population size X. This food improves the physiological state S and increases reproduction

$$\frac{dR}{dt} = I - R(pX + c) \tag{E7.7.1}$$

$$\frac{dS}{dt} = pR - mS \tag{E7.7.2}$$

$$\frac{dX}{dt} = rX(S - S_0) \tag{E7.7.3}$$

where I is the rate of input R to the system, p is the harvesting rate by X, c is the rate of removal R for reasons other than consumption by X, m is the rate of metabolic breakdown of S, S_0 is the threshold S at which reproduction exceeds mortality, and r is the reproductive rate of X based on metabolic intake.

Find the covariation pattern between the resource R and its consumer X.

Solution

From direct time averaging of the Equations E7.7.1, 2, and 3

$$\bar{S} = S_0 \tag{E7.7.4}$$

$$p\bar{R} = m\bar{S} \tag{E7.7.5}$$

$$\text{cov}(X, S) = 0 \tag{E7.7.6}$$

$$\text{cov}(R, S) > 0 \tag{E7.7.7}$$

But cov (R, X) is not given directly. Consider however the derivative of the multiplication of X with S

$$\frac{d(XS)}{dt} = X\frac{dS}{dt} + S\frac{dX}{dt} \quad \text{or}$$

$$\frac{d(XS)}{dt} = pRX - mSX + rXS(S - S_0) \quad \text{(E7.7.8)}$$

The last term is $rX(S - S_0 + S_0)(S - S_0)$, since the addition and subtraction of S_0 does not alter it, the expected value of which is

$$E\left[S\frac{dX}{dt}\right] = rE[X(S - S_0)^2] + rS_0 E[X(S - S_0)] \quad \text{(E7.7.9)}$$

which is positive since X and $(S - S_0)^2$ are both positive while $rS_0 E[X(S - S_0)]$ can be found from the expansion of terms $E[XS] - S_0 E[X]$: $\bar{X}\bar{S} + \text{cov}(X, S) - \bar{X}S_0$. From E7.7.4 and E7.7.5 this is zero. Therefore the only remaining term in E7.7.8 yields

$$p\,\text{cov}\,(R, X) - m\,\text{cov}\,(S, X) < 0$$

Equation E7.7.8 has sustained bounded motion since the consumer and its own metabolism neither grows infinitely large or vanishes, thus

$$0 = p\bar{R}\bar{X} + p\,\text{cov}\,(R, X) - m\bar{S}\bar{X} - m\,\text{cov}\,(S, X) + rE[X(S - S_0)^2]$$

Substituting E7.7.5 and E7.7.6 along with the property of the expected value of this last term gives

$$-rp\,\text{cov}\,(R, X) = rE[X(S - S_0)^2] > 0$$

so that

$$\text{cov}\,(R, X) < 0$$

Notice that the equation for R was not used, and the relation between the covariance of R and X would hold for any (dR/dt) compatible with sustained bounded motion. □

We can multiply Equations E7.7.1, 2, and 3 by the derivative before averaging. Following the procedure of the R, S, X system of Example 7.7

$$\left(\frac{dS}{dt}\right)\left(\frac{dS}{dt}\right) = (pR - mS)\frac{dS}{dt}$$

which can be written as

$$\left(\frac{dS}{dt}\right)^2 = pR\frac{dS}{dt} - mS\frac{dS}{dt}$$

The expected value of $(dS/dt)^2$ is obviously positive since

$$E\left[\left(\frac{dS}{dt}\right)^2\right] = E\left[\frac{dS}{dt}\right]E\left[\frac{dS}{dt}\right] + \text{var}\left(\frac{dS}{dt}\right)$$

$$= 0 + \text{var}\left(\frac{dS}{dt}\right)$$

But, as we found in Example 7.7

$$E\left[S\frac{dS}{dt}\right] = 0$$

since it is the derivative of $\frac{1}{2}S^2$ and the time average of the derivative of any bounded function is zero. Therefore

$$E\left[R\frac{dS}{dt}\right] > 0$$

Substituting the value of dS/dt, we have

$$E[R(pR - mS)] > 0$$

Expanding the expectation operation over the terms (and using the σ^2 notation to signify variance)

$$p\sigma_R^2 - m\,\text{cov}(R, S) > 0$$

Multiplying (dS/dt) by S in E.7.7.2, we get

$$p\,\text{cov}(R, S) = m\sigma_S^2$$

Therefore

$$p^2\sigma_r^2 > m^2\sigma_S^2$$

Note that this holds regardless of the equations for R and X, whenever we have SBM. Even when X or R are driven by time-dependent variables, which may have any kind of bounded behavior, a single equation already establishes the relation between the variances of R and S.

Up to now we have emphasized inferences about the relations among the variables themselves. Sometimes we want the relations between variables and rates of change. We already saw that $E[X, (dX/dt)] = 0$. By multiplying by (dX/dt) before averaging we can find some of these.

Suppose that
$$\frac{dX}{dt} = X(pR - \theta)$$
where R is another variable and θ is constant. First divide by X and multiply by (dX/dt)
$$\left(\frac{dX}{dt}\right)^2 \frac{1}{X} = (pR - \theta)\frac{dX}{dt} \qquad (7.44)$$

Since X is always positive (a population with SBM), the left side is positive. The average of $\theta(dX/dt)$ is zero, so we are left with
$$E\left[R\frac{dX}{dt}\right] > 0$$

If instead we had
$$\frac{dX}{dt} = X(pR - f(X))$$
the same argument would hold since
$$E\left[f(X)\frac{dX}{dt}\right] = 0$$

What about $E[X(dR/dt)]$? We already have seen (Equation 7.24) that
$$E\left[X\frac{dY}{dt}\right] + E\left[Y\frac{dX}{dt}\right] = 0$$

Therefore
$$E\left[X\frac{dR}{dt}\right] < 0 \qquad (7.45)$$

The equation for R may be of any form, no matter how unmanageable mathematically. We might not be able to find $E[X(dR/dt)]$ directly. Or, dR/dt may be a function of other variables but not X. The relation 7.45 still would hold.

But if θ is a time-dependent parameter or depends on other variables in the system, the relation no longer is necessary. Therefore from 7.44 we would have only
$$pE\left[R\frac{dX}{dt}\right] - E\left[\theta\frac{dX}{dt}\right] > 0$$

This procedure also can be applied to the system of Example 7.7. Multiply Equation E7.7.2 by (dS/dt) and take expected values. Since

$$E\left[R\frac{dS}{dt}\right] > 0$$

then

$$E\left[S\frac{dR}{dt}\right] < 0$$

But

$$S\frac{dR}{dt} = S[I - R(pX + c)]$$

Therefore cov $[S, R(pX + c)] > 0$.

Since $R(pX + c)$ is the removal of the resource R per unit time, it means that the animals are in top physiological condition when R is flowing quickly through the system.

In the previous operation we had inferences about the left side of the equation which then could be used to interpret the right side. Now we do the opposite. Consider the familiar pair of equations

$$\frac{dR}{dt} = I(t) - R(pX + c) \tag{7.46}$$

$$\frac{dX}{dt} = X(pR - \theta) \tag{7.47}$$

Multiply Equation 7.46 by $pR - \theta$

$$(pR - \theta)\frac{dX}{dt} = X(pR - \theta)^2$$

Since the right side is non-negative and positive, except for a stable equilibrium

$$E\left[(pR - \theta)\frac{dX}{dt}\right] > 0$$

Expanding for clarity shows that we have

$$pE\left[R\frac{dX}{dt}\right] - E\left[\theta\frac{dX}{dt}\right] > 0 \tag{7.48}$$

Since $E[\theta(dX/dt)] = 0$, substitute into Equation 7.48 to find

$$pE\left[R\frac{dX}{dt}\right] > 0 \tag{7.49}$$

Recalling Equation 7.24

$$E\left[\frac{d(RX)}{dt}\right] = E\left[R\frac{dX}{dt} + X\frac{dR}{dt}\right] = 0$$

and given 7.49, we conclude that

$$E\left[X\frac{dR}{dt}\right] < 0 \tag{7.50}$$

Therefore multiply Equation 7.46 by X and average, using the results of 7.50, to get

$$\text{cov}(X, I) - p\,\text{cov}(X, RX) - c\,\text{cov}(R, X) < 0$$

But we already know from 7.47 that $\text{cov}(R, X) = 0$. Multiply Equation 7.47 by X and take the expected value to get

$$p\,\text{cov}(X, RX) = \theta\,\text{var}(X)$$

Thus

$$\text{cov}(X, I) - \theta\sigma_X^2 < 0$$

and finally either

$$\text{cov}(X, I) < 0 \quad \text{or}$$

$$\sigma_X^2 > \frac{p^2\sigma_I^2}{\theta^2}$$

EXAMPLE 7.8

Consider the resource (R) consumer (X) pair of Equations E7.8.1 and 2.

$$\frac{dR}{dt} = I - R(pX + c) \tag{E7.8.1}$$

$$\frac{dX}{dt} = X(pR - \theta) \tag{E7.8.2}$$

Suppose that the uptake rate $p = p(t)$ is not a constant. It may be a function of temperature or other conditions. Describe the correlation of the consumer pressure, pX, on R and interpret the results in terms of the turnover rate of the resource. (Recall that by turnover we mean here the inverse of residence time—how long does any single unit of resource R remain in the system.)

Solution

From Equation E7.8.1 we take averages both before and after dividing by X to get

$$\text{cov}(X, pR) = 0$$

Likewise for (E7.8.2), we take the expected value before and after dividing by R to find

$$IE\left[\frac{1}{R}\right] = E[pX] + c \quad \text{and} \tag{E7.8.3}$$

$$I\left(\frac{1}{\bar{R}}\right) = E[pX] + c + \frac{\text{cov}(R, pX)}{\bar{R}} \tag{E7.8.4}$$

Since $E[1/R] > 1/E[R]$ means that for (E7.8.3) to be greater than (E7.8.4) it must be that

$$\text{cov}(R, pX) < 0$$

Because pX is the consumer pressure on R, the negative correlation means that when R is abundant the turnover rate is slow and when R is low the turnover rate is rapid. The zero correlation of X with pR means that X is uncorrelated with its effective resource uptake. But the correlation of X with R is unknown. □

Finite Intervals

Up until now we have used the approximation

$$E\left\{\frac{dX}{dt}\right\} = \frac{1}{t}[X(t) - X_0] = 0$$

We let t get very large so that in the limit the expected value is zero. If the time interval of observation is short, however, the approximation has to be replaced by the exact relation

$$E\left\{\frac{dX}{dt}\right\} = \frac{1}{t}(X(t) - X_0) \tag{7.51}$$

and

$$E\left\{\frac{1}{X}\frac{dX}{dt}\right\} = \frac{1}{t}\ln\frac{X(t)}{X_0} \tag{7.52}$$

The finite time interval introduces an additional term into the expected values of equations and therefore could alter the sign of the correlations between variables. Furthermore, over a finite interval it is not necessary that the system have sustained bounded motion as a limiting behavior.

Consider

$$\frac{dX}{dt} = rX \tag{7.53}$$

where r may be any function of time. Then, taking average values,

$$\frac{1}{t}(X_t - X_0) = \bar{r}\bar{X} + \text{cov}(r, X) \tag{7.54}$$

Divide 7.53 by X and then take expected values to get

$$\frac{1}{t} \ln \frac{X_t}{X_0} = \bar{r} \tag{7.55}$$

Substituting this value of $\bar{r}$ into 7.54, we find

$$\text{cov}(r, X) = \frac{1}{t}\left[X_t - X_0 - \bar{X} \ln \frac{X_t}{X_0}\right] \tag{7.56}$$

If r were constant, then Equation 7.53 would have the solution

$$X(t) = X_0 e^{rt} \tag{7.57}$$

and substituting 7.57 for X_t in the expected value term $1/t(X_t - X_0)$ yields

$$\frac{1}{t}(X_t - X_0) = \frac{X_0}{t}(e^{rt} - 1) \tag{7.58}$$

Simplifying 7.58 yields

$$\frac{1}{t} \ln \frac{X_t}{X_0} = r \tag{7.59}$$

The average value of X is

$$\bar{X} = \frac{1}{t} \int_0^t X_0 e^{r\tau} \, d\tau \tag{7.60}$$

$$\bar{X} = \frac{X_0}{t} \int_0^t e^{r\tau} \, d\tau$$

$$\bar{X} = \frac{X_0}{rt}(e^{rt} - e^{r \cdot 0})$$

which is

$$\bar{X} = \frac{X_0}{rt}(e^{rt} - 1) \tag{7.61}$$

Therefore, the right-hand side of Equation 7.56 can be found with substitution of 7.59 and 7.61 to be

$$\frac{1}{t}\left\{X_t - X_0 - \bar{X}\ln\frac{X_t}{X_0}\right\} = 0 \tag{7.62}$$

But if $r(t)$ varies in such a way that r is on the average $> \bar{r}$ at first and later is below $\bar{r}$, then

$$\bar{X} > \frac{1}{t}\int_0^t X_0 e^{\bar{r}\tau}\, d\tau \tag{7.63}$$

and, looking back at 7.56, we find

$$\text{cov}(r, X) < 0 \tag{7.64}$$

That is, even though large r increases X, there can be negative covariance.

EXAMPLE 7.9

Let

$$\frac{dX}{dt} = X[K(t) - X] \tag{E7.9.1}$$

Find the covariance $\text{cov}(K, X)$ over a finite time interval.

Solution

The expected value of E7.9.1 is

$$E\left[\frac{dX}{dt}\right] = \frac{1}{t}[X(t) - X_0] = \bar{X}[\bar{K} - \bar{X}] - \text{var}(X) - \text{cov}(X, K) \tag{E7.9.2}$$

If we divide E7.9.1 by X before averaging, we get

$$E\left[\frac{1}{X}\frac{dX}{dt}\right] = \frac{1}{t}\ln\frac{X(t)}{X_0} \tag{E7.9.3}$$

So that

$$\bar{K} - \bar{X} = \frac{1}{t}\ln\frac{X(t)}{X_0} \tag{E7.9.4}$$

Substituting E7.9.4 into E7.9.2 gives

$$E\left[\frac{dX}{dt}\right] = \frac{1}{t}[X(t) - X_0] = \frac{\bar{X}}{t}\ln\frac{X(t)}{X_0} + \text{cov}(K, X) - \text{var}(X) \tag{E7.9.5}$$

and

$$\text{cov}(K, X) = \text{var}(X) + \frac{1}{t}[X(t) - X_0] - \frac{\bar{X}}{t}\ln\frac{X(t)}{X_0} \tag{E7.9.6}$$

The first term on the right shows that in the limit as t increases, the covariance is positive. But in the finite sequence, if $\bar{X}\ln(X(t))/(X_0) > [X(t) - X_0]$, it is possible to get a negative covariance. In order to determine when this may happen, note that if X first increases very rapidly toward $X(t)$ and then slows down, $\bar{X}$ can be close to $X(t)$. But if X remains near X_0 for a long time and then changes rapidly to $X(t)$, $\bar{X}$ can be close to X_0. Suppose that X changes exponentially

$$\frac{dX}{dt} = rX \quad \text{and} \tag{E7.9.7}$$

$$X(t) = X_0 e^{rt} \tag{E7.9.8}$$

Then

$$X(t) - X_0 = X_0[e^{rt} - 1]$$

and from E.7.9.8,

$$\ln\frac{X(t)}{X_0} = rt$$

and

$$\bar{X} = \frac{1}{t}\int_0^t X(\tau)\,d\tau \quad \text{or}$$

$$\bar{X} = \frac{X_0}{t}\int_0^t e^{r\tau}\,d\tau$$

$$= \frac{X_0}{rt}[e^{rt} - 1]$$

Finally

$$\frac{1}{t}[X(t) - X_0] - \frac{\bar{X}}{t}\ln\frac{X(t)}{X_0} = \frac{X_0}{t}[e^{rt} - 1] - \frac{X_0}{t}[e^{rt} - 1] \tag{E7.9.9}$$

which is zero. Therefore, if X increases from X_0 to $X(t)$ more rapidly than exponentially at first, $\bar{X}$ will be large and the time-dependent term will be negative. Further, if X increases rapidly and then remains near $X(t)$, var(X) will be small. Under these conditions, cov(K, X) can be negative, although K increases X. □

EXAMPLE 7.10
Find the covariance of X and Y when

$$\frac{dX}{dt} = X(K(t) - X - pY) \tag{E7.10.1}$$

$$\frac{dY}{dt} = Y(pX - \theta) \tag{E7.10.2}$$

Solution
In the limit, cov$(X, Y) = 0$ and cov$(X, K) = $ var$(X) > 0$. In finite time, applying the previous results to Equation E7.10.2

$$\text{cov}(X, Y) = \frac{1}{t}\left[Y_t - Y_0 - \bar{Y}\ln\frac{Y_t}{Y_0}\right] \tag{E7.10.3}$$

If Y changed at a uniform exponential rate ($pX - \theta$ would be constant for this to happen), then cov(X, Y) would be zero.

But if Y first increases faster than the best-fit exponential and then increases more slowly, cov$(X, Y) < 0$. Similarly, if Y decreases rapidly at first and then more slowly, cov$(X, Y) < 0$. But if change is slow at first and then accelerates, cov$(X, Y) > 0$. □

The next three examples all examine the same pair of equations.

$$\frac{dR}{dt} = R[K - R - pA] \tag{7.65}$$

$$\frac{dA}{dt} = A[pR - \theta] \tag{7.66}$$

where K is the carrying capacity for R. A consumes R at rate p, and θ is the death rate of A. The loop diagram for these equations is shown in Figure 7.3.

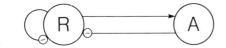

Fig. 7.3

EXAMPLE 7.11

Determine what the correlation pattern will be among the variables if there is an abrupt change in the carrying capacity.

Solution

Suppose that K increases from its original value K_0 to a new, greater K_1. From the loop diagram or from Equation 7.66 we see that the equilibrium value of R does not change

$$R^* = \frac{\theta}{p} \tag{E7.11.1}$$

The immediate effect of a jump in K, however, is to increase R. Therefore we can trace R from some $R_0 > \bar{R}$ a short time after K shifted, up to some peak $R_{\max}$ and back to R_0 before eventually approaching $\bar{R}$ again. Since $R_0 = R_t$, $E(dR/dt) = 0$ and

$$\bar{R}[K - R - p\bar{A}] - \text{var}(R) - p \,\text{cov}(R, A) = 0 \tag{E7.11.2}$$

Similarly, from $E((1/R)(dR/dt)) = 0$

$$K - \bar{R} - p\bar{A} = 0 \tag{E7.11.3}$$

during the transition. Therefore, from (E7.11.2) we have

$$p \,\text{cov}(R, A) = -\text{var}(R) < 0 \tag{E7.11.4}$$

That is, during the transition to the new equilibrium, R and A are negatively correlated over any interval for which the initial and final values of R are equal.

From Equation 7.66 we find

$$\frac{A_t - A_0}{t} = \bar{A}(p\bar{R} - \theta) + p \,\text{cov}(A, R) \tag{E7.11.5}$$

and

$$\frac{1}{t} \ln \frac{A_t}{A_0} = p\bar{R} - \theta, \quad \text{and} \tag{E7.11.6}$$

$$\bar{R} = \frac{\theta}{p} + \frac{1}{tp} \ln \frac{A_t}{A} \tag{E7.11.7}$$

Since R is always greater than θ/p during the period of observation, $A_t > A_0$ and $p\bar{R} - \theta > 0$. That is, A has increased. Substituting the value of $\bar{R}$ from E7.11.7 into E7.11.3, we find

$$\bar{A} = \frac{\left(K_1 - \dfrac{\theta}{p}\right)}{p} - \frac{1}{tp^2} \ln \frac{A_t}{A_0} \tag{E7.11.8}$$

relating the average $\bar{A}$ during this period to the starting and ending points, the elapsed time, and the new equilibrium value, $(K_1 - \theta/p)/p$. □

EXAMPLE 7.12

Imagine that A is a pest that occasionally erupts to plague levels. Consider what happens to A and R when, due to some temporary circumstances in the environment, or because of immigration, there is a sudden increase in the abundance of A.

Solution

After some transient changes the previous equilibrium will be restored. But the immediate effect of the increase in A is to reduce R. Therefore let us follow R from some $R_0 < \bar{R}$ down to a minimum R and back to R_0 on the way toward $\bar{R}$ again. The procedures of the previous example can be used here as well, so that cov$(A, R) < 0$. But this time A decreases as the outbreak wanes. Therefore $A_t < A_0$ and $p\bar{R} - \theta < 0$. Combining Equations E7.11.5 and E7.11.7 of the previous example, we have

$$\frac{1}{t}\left(A_t - A_0 - A^* \ln \frac{A_t}{A_0}\right) < 0 \tag{E7.12.1}$$

Therefore A decreases more rapidly than exponentially at first and then slows down. □

EXAMPLE 7.13

Follow the trajectories of R and A if there is a sudden decrease in R.

Solution

R will return toward equilibrium, but first A will decrease due to its loss of food, and then A will come back to it toward its equilibrium. We can follow A from some $A_0 < A^*$ down to a minimum and then back through A_0 on the way to equilibrium again. Therefore $A_t = A_0$ and the expected value of Equation 7.66 gives

$$\bar{A}(p\bar{R} - \theta) + p \operatorname{cov}(A, R) = 0 \quad \text{and} \tag{E7.13.1}$$

$$p\bar{R} - \theta = 0 \tag{E7.13.2}$$

Therefore $\operatorname{cov}(A, R) = 0$ over this transition. The fact that $p\bar{R} - \theta = 0$ means that since R was originally reduced below equilibrium, R^*, it must overshoot above $\bar{R}$ before returning toward equilibrium at R^*.

From Equation 7.66 we can find that

$$\operatorname{var}(R) = R^*(K - R^* - p\bar{A}) + \frac{1}{t} \ln \frac{R_0}{R_t} \tag{E7.13.3}$$

Since $\bar{A} > A^*$ during the transition

$$\operatorname{var}(R) < \frac{1}{t} \ln \frac{R_0}{R_t} \tag{E7.13.4}$$

which relates the variance of the transition to the magnitude of the initial displacement and the elapsed time. $\square$

Discrete-Time Processes

The argument which gave us $E(dX/dt) = 0$ for a process with sustained bounded motion applies also to difference equations of the form given in Equation 7.67.

$$X(t + \Delta t) - X(t) = \Delta X \quad \text{or} \tag{7.67}$$

$$X(t + \Delta t) = X(t) + f(X_1, X_2, \cdots, X_n) \Delta t \tag{7.68}$$

from which we find

$$E\left\{\frac{\Delta X}{\Delta t}\right\} = E\{f(X_1, X_2, \cdots, X_n)\} = 0 \tag{7.69}$$

But

$$\Delta(X^2) = (X + \Delta X)^2 - X^2 \tag{7.70}$$

which is

$$2(\Delta X)X + (\Delta X)^2 \tag{7.71}$$

Therefore, when we cannot ignore $(\Delta X)^2$

$$E\left\{X \frac{\Delta X}{\Delta t}\right\} < 0 \tag{7.72}$$

That is, a variable and its discrete rate of change have a negative correlation. Similarly

$$\Delta(XY) = (X + \Delta X)(Y + \Delta Y) - XY \tag{7.73}$$

so that

$$E\left\{X\frac{\Delta Y}{\Delta t} + Y\frac{\Delta X}{\Delta t} + \frac{\Delta X}{\Delta t}\frac{\Delta Y}{\Delta t}\right\} = 0 \quad \text{and} \tag{7.74}$$

$$E\left\{X\frac{\Delta Y}{\Delta t} + Y\frac{\Delta X}{\Delta t}\right\} = -E\left\{\frac{\Delta X}{\Delta t}\frac{\Delta Y}{\Delta t}\right\} \tag{7.75}$$

instead of zero.

These approximations may be important for short sequences of observations.

Time Averaging and Loop Analysis

Our objective is to find out which results of loop analysis also hold when the assumption of moving equilibrium is relaxed.

First consider a system of linear equations

$$\frac{dX_i}{dt} = S_i(t) + \sum a_{ij}X_j \tag{7.76}$$

and take the expected values. This gives

$$\sum a_{ij}\bar{X}_j = -\bar{S}_i \tag{7.77}$$

Equation 7.11 is of the same form as Equations 5.3 and A.28 (in the Appendix)—except that X_j^*, the equilibrium value, has been replaced by the average value $\bar{X}_j$, in the evaluation of a_{ij} according to (A.21). Therefore, *if a system of linear equations shows sustained bounded motion due to added forcing parameters $S_i(t)$, the loop analysis results for the change in equilibrium value X_j^* due to parameter change also applies exactly for changes in average value, $\bar{X}_j$.*

In common models of ecology, the equations are more often of the form

$$\frac{dX_i}{dt} = X_i[S_i(t) + a_{ij}X_{ij}] \tag{7.78}$$

Here we can divide by X_i provided it is bounded away from zero, and then take expected values to get Equation 7.77 again. This also would hold if the equation had the form

$$\frac{dX_i}{dt} = g(X_i)[S_i(t) + \sum a_{ij}X_j] \tag{7.79}$$

provided $g(X_i)$ is never zero. These equations are referred to as exactly linearizable since we always can get linear equations in the expected values by dividing by $g(X_i)$ before averaging. Therefore,

If dX_i/dt are a set of exactly linearizable equations, then the results of loop analysis for the change in X_i^* in response to parameter change apply exactly to $\bar{X}_i$.

EXAMPLE 7.14

Return to the examples of Chapter 4. Assume that the equations are exactly linearizable, and interpret the results of the loop analysis. □

Linear Equations

Suppose we have a system of linear equations

$$\frac{dx_i}{dt} = s_i(t) + \sum a_{ij}x_j \tag{7.80}$$

Then, taking expected values we obtain the corresponding equations

$$\bar{s}_i + \sum a_{ij}\bar{x}_j = 0$$

Differentiate 7.80 with respect to some parameter c which may be $\bar{s}_i$ or any of the a_{ij}. Then

$$\sum a_{ij}\frac{\partial \bar{x}_j}{\partial c} = -\frac{\partial f_i}{\partial c} \tag{7.81}$$

This is analogous to the equation in the Appendix for which the loop analysis algorithm was derived, except that the unknowns are for the average values of the variable, $(\partial \bar{x}_i/\partial c)$, instead of the equilibrium values, $(\partial x^*/\partial c)$. Therefore the same procedure of the loop analysis can be used and

$$\frac{\partial \bar{x}_i}{\partial c} = \sum \frac{\frac{\partial f_i}{\partial c} P_{ij} F^{(\text{complement})}}{F_n} \tag{7.82}$$

Furthermore, if f_i is linearizable

$$\frac{dx_i}{dt} = g_i(x_i)\{s_i(t) + \sum a_{ij}x_j\}$$

As long as $g_i(x_i)$ is bounded we can divide by $g_i(x_i)$ and take expected values to get Equation 7.90 again.

Therefore, if a set of differential equations can be linearized, loop analysis gives the direction of change of the mean value of variables when parameters are changed.

Nonlinear Equations

If the equations cannot be linearized, the operation of taking expected values introduces higher-order moments in addition to averages. Thus if

$$\frac{dx}{dt} = x[s(t) - x - x^2] \qquad (7.83)$$

dividing by x and using the expectation operator gives

$$\bar{s} - \bar{x} - \bar{x}^2 - \operatorname{var}(x) = 0 \qquad (7.84)$$

This gives us information both about mean and variances. First we pursue inferences about the mean values, then about variances and covariances.

Differentiate 7.84 with respect to $\bar{s}$

$$\frac{\partial \bar{x}}{\partial \bar{s}}(1 + 2\bar{x}) + \frac{\partial \operatorname{var}(x)}{\partial \bar{s}} = 1 \qquad (7.85)$$

Immediately we encounter a difficulty. There is only a single equation, but two unknowns ($\bar{x}$ and $\operatorname{var}(x)$). Therefore we can solve for $(\partial \bar{x}/\partial \bar{s})$, but in terms of $(\partial \operatorname{var}(x)/\partial \bar{s})$:

$$\frac{\partial \bar{x}}{\partial \bar{s}} = \frac{1 - \dfrac{\partial \operatorname{var}(x)}{\partial \bar{s}}}{1 + 2\bar{x}}$$

This is not very useful. It does show that for big changes in variance ($(\partial \operatorname{var}(x)/\partial \bar{s}) > 1$) the mean and variance change in opposite directions, and that for very large $\bar{x}$ the mean changes much less than the variance. It is more important to note, however, that the variance behaves as if it were part of the input to $\bar{x}$ along with $\bar{s}(t)$. This holds in general, as will be seen below. *Nonlinear terms give variances, covariances, and other higher-order moments which behave in the equations as if they were inputs to the variable in whose equation they appear.*

In Equation 7.84, $\bar{s}$ appears as a parameter of the expected value equation so that when we differentiate with respect to $\bar{s}$ there is an input equal to $(d\bar{s}/d\bar{s}) = 1$. Suppose instead that we asked, what is the effect of changing the variance of $s(t)$ on the mean of x? Since σ_s^2 does not appear at all in the expected value equation, the derivative with respect to σ_s^2 is zero. Therefore, if we had linear equations, as in 7.81, all the $(\partial \bar{x}_i/\partial \sigma_x^2)$ would be equal to zero: The averages of the variables depend only on the mean of the environmental parameter, $s_i(t)$, and not its variance.

But with the nonlinear expressions, something new happens. Refer back to Equation 7.84 and differentiate with respect to the variance of s.

$$\frac{\partial \bar{x}}{\partial \sigma_s^2}(1 + 2\bar{x}) + \frac{\partial \operatorname{var}(x)}{\partial \sigma_s^2} = 0$$

and
$$\frac{\partial \bar{x}}{\partial \sigma_s^2} = -\frac{\dfrac{\partial \operatorname{var}(x)}{\partial \sigma_s^2}}{(1 + 2\bar{x})}$$

The average value of x is sensitive to changes in the variance of the environment, and changes in the opposite direction from the variance. Further, their relative sensitivities depend on the denominator $1 + 2\bar{x}$. As $\bar{x}$ increases, an increasing proportion of the impact of changed σ_s^2 is absorbed by σ_x^2 as against $\bar{x}$. The denominator term $1 + 2\bar{x}$ is equivalent to the F_n (feedback of the whole system) except that F_n is evaluated at equilibrium rather than at the average values.

We now can generalize these results further. Suppose

$$\frac{dx_j}{dt} = f_j(x_1, x_2, x_3, \cdots, c_1, c_2, \cdots) \tag{7.86}$$

is an arbitrary function for the x's, but still linear in the c's, the parameters. We can expand f_j in a Taylor series around the mean values (see Appendix) x_j to obtain

$$\begin{aligned}
\frac{dx_j}{dt} = {} & f_j(\bar{x}_1, \bar{x}_2, \bar{x}_3, \cdots; \bar{c}_1, \bar{c}_2, \bar{c}_3, \cdots) \\
& + \sum_i \frac{\partial f_j(\bar{x}_1, \bar{x}_2, \bar{x}_3, \cdots; \bar{c}_1, \bar{c}_2, \bar{c}_3, \cdots)}{\partial x_i}(x_i - \bar{x}_i) \\
& + \sum_h \frac{\partial f_j(\bar{x}_1, \bar{x}_2, \bar{x}_3, \cdots; \bar{c}_1, \bar{c}_2, \bar{c}_3, \cdots)}{\partial c_h}(c_h - \bar{c}_h) \\
& + \frac{1}{2}\sum_i \frac{\partial^2 f_j(\bar{x}_1, \bar{x}_2, \bar{x}_3, \cdots; \bar{c}_1, \bar{c}_2, \bar{c}_3, \cdots)}{\partial x_i^2}(x_i - \bar{x}_i)^2 \\
& + \sum_{i,k} \frac{\partial^2 f_j(\bar{x}_1, \bar{x}_2, \bar{x}_3, \cdots; \bar{c}_1, \bar{c}_2, \bar{c}_3, \cdots)}{\partial x_i \partial x_k}(x_i - \bar{x}_i)(x_k - \bar{x}_k) \\
& + \sum_{h,i} \frac{\partial^2 f_j(\bar{x}_1, \bar{x}_2, \bar{x}_3, \cdots; \bar{c}_1, \bar{c}_2, \bar{c}_3, \cdots)}{\partial c_h \partial x_i}(c_h - \bar{c}_h)(x_i - \bar{x}_i) \\
& + \frac{1}{2}\sum_i \frac{\partial^2 f_j(\bar{x}_1, \bar{x}_2, \bar{x}_3, \cdots; \bar{c}_1, \bar{c}_2, \bar{c}_3, \cdots)}{\partial c_h^2}(c_h - \bar{c}_h)^2 \\
& + \sum_{h,k} \frac{\partial^2 f_j(\bar{x}_1, \bar{x}_2, \bar{x}_3, \cdots; \bar{c}_1, \bar{c}_2, \bar{c}_3, \cdots)}{\partial c_h \partial c_k}(c_h - \bar{c}_h)(c_k - \bar{c}_k) \\
& + \text{higher-order terms} \tag{7.87}
\end{aligned}$$

We condense the notation to omit the arguments of the functions, the x's and c's, which are taken at their average values, and take the expected value of the whole expression of Equation 7.87.

$$0 = f_j + \frac{1}{2}\sum_i \frac{\partial^2 f_j}{\partial x_i^2}\operatorname{var}(x_i) + \sum_{i,k}\frac{\partial^2 f_j}{\partial x_i \, \partial x_k}\operatorname{cov}(x_i, x_k)$$

$$+ \sum_{h,i}\frac{\partial^2 f_j}{\partial x_i \, \partial c_h}\operatorname{cov}(x_i, c_h) + \sum_{h,k}\frac{\partial^2 f_j}{\partial c_h \, \partial c_k}\operatorname{cov}(c_h, c_k)$$

$$+ \frac{1}{2}\sum_h \frac{\partial^2 f_j}{\partial c_h^2}\operatorname{var}(c_h) + \text{higher moments} \tag{7.88}$$

Differentiate (7.88) with respect to some parameter, c.

$$0 = \frac{\partial f_j}{\partial c} + \sum_i \frac{\partial f_j}{\partial \bar{x}_i}\frac{\partial \bar{x}_i}{\partial c} + \frac{1}{2}\sum_i \frac{\partial^2 f_j}{\partial x_i^2}\frac{\partial \operatorname{var}(x_i)}{\partial c} + \sum_{i,k}\frac{\partial^2 f_j}{\partial x_i \, \partial x_k}\frac{\partial \operatorname{cov}(x_i, x_k)}{\partial c}$$

$$- \sum_{h,i}\frac{\partial^2 f_j}{\partial x_i \, \partial c_h}\frac{\partial \operatorname{cov}(x_i, c_h)}{\partial c} + \sum_{h,k}\frac{\partial^2 f_j}{\partial c_h \, \partial c_k}\frac{\partial \operatorname{cov}(c_h, c_k)}{\partial c}$$

$$+ \frac{1}{2}\sum_h \frac{\partial^2 f_j}{\partial c_h^2}\frac{\partial \operatorname{var}(c_h)}{\partial c} + \frac{1}{2}\sum_i \frac{\partial^3 f_j}{\partial x_i^2 \, \partial c}\operatorname{var}(x_i)$$

$$+ \text{ other terms in the third and higher derivatives}$$
$$\text{and higher moments} \tag{7.89}$$

The first two terms in Equation 7.89 are the same as in the loop formula in the Appendix. The additional terms introduce new unknowns, the sensitivities of variances and covariances to parameter change. Since we have more unknowns than equations we cannot solve for the new unknowns. Instead we add them to $\partial f_i/\partial c$ as part of the "input" term for loop analysis. All such terms can be lumped as a single term U coming from nonlinear sustained bounded motion. But for purposes of illustration we also can simplify by assuming that the functions f are at most quadratic in the x's and that the parameter c enters most as a linear term. Now we can apply the loop analysis algorithm to solve for the $\partial \bar{x}_i/\partial c$ as given by Equation 7.90.

For the moment we will assume that the functions f_i are quadratic so that all third derivatives and higher are zero. Now we can apply the loop analysis algorithm to solve for the $(\partial \bar{x}_i/\partial c_h)$

$$\frac{\partial \bar{x}_i}{\partial c_h} = \frac{\sum\left[\dfrac{\partial f_j}{\partial c_h} + \dfrac{1}{2}\dfrac{\partial^2 f_j}{\partial \bar{x}_i^2}\dfrac{\partial \operatorname{var}(x_i)}{\partial c_h} + \sum \dfrac{\partial^2 f_j}{\partial \bar{x}_i \bar{x}_k}\dfrac{\partial \operatorname{cov}(x_i, x_k)}{\partial c_h}\right] \times p_{ij}F^{(\text{comp})}}{F_n}$$

$$\tag{7.90}$$

TIME AVERAGING   229

Or, using the σ_i^2 and σ_{ij} notation for variance and covariance, respectively, the formula is

$$\frac{\partial \bar{x}_i}{\partial c_h} = \frac{\sum \left[\frac{\partial f_j}{\partial c_h} + \frac{1}{2}\frac{\partial^2 f_j}{\partial \bar{x}_i^2}\frac{\partial \sigma_i^2}{\partial c_h} + \sum \frac{\partial^2 f_j}{\partial \bar{x}_i \partial \bar{x}_k}\frac{\partial \sigma_{ik}}{\partial c_h}\right] \times p_{ij} F^{(comp)}}{F_n} \quad (7.91)$$

This differs from the loop result in two ways: The coefficients of the path, complement, and denominator are the $(\partial f_j / \partial \bar{x}_i)$ evaluated at average values; and the input $(\partial f_j / \partial c_h)$ has been expanded to include the unknown sensitivities of the variances and covariances to parameter change.

EXAMPLE 7.15

Given

$$\frac{dx}{dt} = x(k - x - y^2) \quad (E7.15.1)$$

$$\frac{dy}{dt} = y(ax - \theta) \quad (E7.15.2)$$

where $k = k(t)$ is a time-dependent parameter. What happens to the variables if the average value of the parameter k, $\bar{k}$, changes? If the variance of k, var (k) changes?

Solution

Divide E7.15.1 and E7.15.2 by x and y, respectively and take averages to get

$$\bar{k} - \bar{x} - \bar{y}^2 - \sigma_y^2 = 0 \quad (E7.15.3)$$

$$a\bar{x} - \theta = 0 \quad (E7.15.4)$$

The loop diagram is shown in Figure 7.4.

Fig. 7.4

The nonlinearity is an input to x. Since its complement, y, has zero feedback

$$\frac{\partial \bar{x}}{\partial \bar{k}} = 0$$

as before. But

$$\frac{\partial \bar{y}}{\partial \bar{k}} = \frac{1 - \frac{\partial \sigma_y^2}{\partial \bar{k}}}{2\bar{y}} \qquad (E7.15.5)$$

This result is not very useful by itself. It does suggest, however, that with a very small $\bar{y}$, $\bar{y}$ is more sensitive to parameter change than var (y), and with large $\bar{y}$ the variance may be relatively more sensitive.

Now what happens when the variance of k changes? Since var (k) does not appear at all in Equation E7.15.3, there is no $(\partial f_i/\partial \text{ var}(k))$. In the linear loop analysis this would be sufficient to show that $\bar{y}$ does not change. But because of the nonlinear term

$$\frac{\partial \bar{y}}{\partial \text{ var}(k)} = \frac{-\frac{\partial \text{ var}(y)}{\partial \text{ var}(k)}}{2\bar{y}} \qquad (E7.15.6)$$

Thus $\bar{y}$ and var (y) change in opposite directions in response to changes in the variance of k (or other statistical measures except $\bar{k}$). In this example the result is obvious from E7.15.3. □

EXAMPLE 7.16

Given the loop model of Figure 7.5, determine the response of the variables in the network due to parameter changes at each node if (a) the consumer A_2 is nonlinear in the equation (dA_2/dt), or (b) the resource nutrient N is nonlinear in its equation.

Solution

Since the graph indicates a stable system there must be some time-dependent parameter driving the fluctuations. If any parameter changes, its effect has two parts: The effect on the average values of the variables as calculated from the rules for moving equilibrium (the loop analysis of Chapter 3), plus the

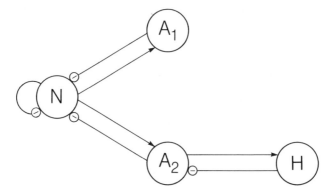

Fig. 7.5

effect of the unknown input from the nonlinearity at A_2. Since A_2 has the satellite H, this nonlinearity will affect only the average $\bar{H}$ but none of the other variables. If the parameter does not appear in the equations for expected values, for example, if it is the variance or skewness of a fluctuating input, then the only changes in average values come from the nonlinearity and effect only $\bar{H}$.

Similarly, a nonlinearity in H affects $\bar{A}_2$ and $\bar{A}_1$ in opposite directions but has no effect on $\bar{N}$ or $\bar{H}$.

Now consider what happens when there is a nonlinearity in H and an input to N. $\bar{A}_1$ will change for both reasons given above and $\bar{A}_2$ will respond only to the nonlinearity. If the nonlinearity is strong enough it will predominate, and A_1 and A_2 will change in opposite directions. *Any input may activate all nonlinear nodes.* □

EXAMPLE 7.17

Holling (1959) introduced the notion of functional response of predator to prey, resulting in various relations between feeding rate and prey abundance. One such relation (Holling Type 2) could give a pair of equations in which the predator is saturated by abundant prey

$$\frac{dx}{dt} = x\left(A - x - \frac{py}{k+x}\right) \tag{E7.17.1}$$

$$\frac{dy}{dt} = y\left(\frac{px}{k+x} - \theta\right) \tag{E7.17.2}$$

Determine the stability conditions and the change in variability to prey x when the input A increases, that is, more prey are introduced.

Fig. 7.6

Solution
Since

$$\frac{\partial\left(\dfrac{dx}{dt}\right)}{\partial x} = -x + \frac{pxy}{(k+x)^2}$$

The system is locally unstable when y is large and x is small. From the loop diagram shown in Figure 7.6, we see that increasing input to x can increase y

without increasing x, until eventually we may have the loop model shown in Figure 7.7.

Fig. 7.7

This system would show SBM. From Equation E7.17.1.

$$\bar{x}(A - \bar{x}) - \sigma_x^2 = p\overline{\bar{y}\left(\frac{x}{k+x}\right)} + p\,\text{cov}\left(y, \frac{x}{k+x}\right) \tag{E7.17.3}$$

But from (E7.17.2), the covariance is zero and

$$p\overline{\left(\frac{x}{k+x}\right)} = \theta$$

Therefore

$$\bar{x}(A - \bar{x}) - \sigma_x^2 = \bar{y}\theta \tag{E7.17.4}$$

Since $\bar{x}$, $\bar{y}$, and σ_x^2 cannot be negative, $\bar{x}$ is always less than A. Therefore $\bar{x}(A - \bar{x}) \leq (A/2)^2$.

If we multiply Equation E7.17.1 by $(k + x)/(x)$, we have

$$\frac{k+x}{x}\frac{dx}{dt} = (A - x)(k + x) - py \tag{E7.17.5}$$

Taking expected values of E7.17.4 and E7.17.5, we get the pair of equations for $\bar{y}$ and σ_x^2 in terms of $\bar{x}$

$$\theta\bar{y} + \sigma_x^2 = \bar{x}(A - \bar{x}) \tag{E7.17.6}$$

$$p\bar{y} + \sigma_x^2 = Ak + (A - k)\bar{x} - \bar{x}^2 \tag{E7.17.7}$$

The solution of the pair of linear Equations E7.17.6 and E7.17.7 is

$$\bar{y} = \frac{\begin{vmatrix} \bar{x}(A - \bar{x}) & 1 \\ Ak + (A - k)\bar{x} - \bar{x}^2 & 1 \end{vmatrix}}{\theta - p}$$

$$\sigma_x^2 = \frac{\begin{vmatrix} \theta & \bar{x}(A - \bar{x}) \\ p & Ak + (A - k)\bar{x} - \bar{x}^2 \end{vmatrix}}{\theta - p} \quad \text{or}$$

$$\sigma_x^2 = \frac{\theta Ak + \theta(A - k)\bar{x} - \theta\bar{x}^2 - p\bar{x}(A - \bar{x})}{\theta - p}$$

and finally
$$\sigma_x^2 = \frac{\theta A k + [(\theta - p)A - \theta k]\bar{x} - (\theta - p)\bar{x}^2}{\theta - p}$$

We do not have the value of $\bar{x}$, but we can set limits to σ_x^2. The above expression reaches a maximum value when
$$\bar{x} = \frac{[(\theta - p)A - \theta k]}{2(\theta - p)}$$

Thus we would have
$$\sigma_x^2 \leq \frac{\left\{\theta A k + \dfrac{[(\theta - p)A - \theta k]^2}{4(\theta - p)}\right\}}{(\theta - p)}$$

This upper bound for σ_x^2 increases with input A. Note that since $x/k + x < 1$ in Equation E7.17.2, p must be greater than θ. □

In summary:

1. For nonequilibrium systems that are linearizable, the loop analysis predictions apply to average values instead of equilibrium values.
2. A nonlinearity behaves like an input to the variable in whose equation it appears.
3. Average values of the variables can change as a result of changes in the variances or other statistical descriptors of the parameters.

Appendix
References
Index

Appendix: Some Mathematical Operations and Concepts

This appendix is intended to aid those unfamiliar with certain terms, mathematical notations, or operations. It cannot substitute for a textbook on any of the topics but should provide enough background to allow everyone to proceed through this primer. The topics included are:

> Summation sign
> Dynamical systems classifications
> Partial derivatives
> Differential equation systems
> Taylor series
> Jacobian matrix
> Changing parameters—the loop formula

Summation Sign

The symbol $\sum_{i=1}^{n} x_i$ signifies that the terms $x_1, x_2, \cdots x_n$ are added together. For example, when $n = 5$, then

$$\sum_{i=1}^{5} x_i = x_1 + x_2 + x_3 + x_4 + x_5$$

A summation is indeterminate when no upper limit is set; hence, $\sum_i x_i$ denotes that all the x's are to be added together, however many there are.

Two summation signs together indicates a *double summation*. Thus

$$\sum_{i=1}^{3} \sum_{j=1}^{4} x_i y_j = x_1 y_1 + x_1 y_2 + x_1 y_3 + x_1 y_4$$
$$+ x_2 y_1 + x_2 y_2 + x_2 y_3 + x_2 y_4$$
$$+ x_3 y_1 + x_3 y_2 + x_3 y_3 + x_3 y_4$$

The double summation is frequently written in the compact form

$$\sum_{i=1, j=1}^{3,4} x_i y_j$$

The order of the variables is not important; the order of the subscripts under the summation sign is. So the indeterminate triple summations listed below are equivalent

$$\sum_{i,j,k} x_i y_j z_k = \sum_{i,j,k} z_k x_i y_j$$

The following are *not* equivalent

$$\sum_{i,j,k} x_i y_j z_k \neq \sum_{k,i,j} x_i y_j z_k$$

Dynamical Systems Classifications

This primer developed from a special subset of dynamical systems theory, determined by a series of dichotomous fields of study within systems theory. In broad terms, systems theory can be segmented as follows:

Linear	:	Nonlinear
Continuous	:	Discrete
Deterministic	:	Stochastic
Nonanticipatory	:	Anticipatory
Time invariant	:	Time varying

The subset of most interest in this book is that of the *linear* and *nonlinear, continuous, deterministic, nonanticipatory,* and *time-invariant* systems. Although this is a special subset, nevertheless many of its theorems apply to other subsets. Some of the general concepts, if not the specific theorems, of this subset are more widely applicable—for example, notions of stability. Digraphs (loop models) can be developed for other subsets of dynamical systems, such as linear, discrete, stochastic, nonanticipatory, and time-invariant systems. (See Lorens, 1964, for engineering applications.)

A brief definition of these terms follows.

LINEAR:NONLINEAR

Linearity is a property of a system when the ouput is proportionally related to the input (the dependent and independent variables, respectively), over the entire range of the input. We recognize an algebraic equation as being linear when it has the generic form for a straight line

$$y = ax + b$$

whereas

$$y = ax + bx^3$$

is clearly not the equation of a straight line. One criterion of linearity is that the *principle of superposition* must apply. Superposition states that results can be added: The output of a system from one input and the output from a second, different input can be added to give the output just as if the sum of both inputs had been impressed upon the system. Bernard Patten (personal communication) has noted that a straight line does not follow the principle of superposition unless $b = 0$, that is, the line passes through the origin. But, of course, any line whose intercept is not zero can be shifted into a new line that does have a zero intercept. This new line satisfies the superposition principle.

CONTINUOUS:DISCRETE

In the systems theory literature a continuous system is one in which the input is defined at all values, with the possible exception of a denumerable set of points. The inputs and outputs vary without breaks except in a finite number of places. In the strict mathematical sense of continuous, any sharp breaks would indicate that in the range of the input that includes the break, the system is discontinuous. We follow the mathematical definition in this book, and do not consider "jumps" in either input or output as being part of a continuous system.

A discrete system has inputs occurring at exact intervals, although these are not necessarily equally spaced.

NONANTICIPATORY:ANTICIPATORY

A nonanticipatory system is one in which the present outcome does not depend on future outcomes.

TIME-INVARIANT:TIME-VARYING

A time-invariant system has an output that remains the same for a given input regardless of the time of input. Given

$$\frac{dy}{dt} = ax + b$$

then for $x = 3$

$$\frac{dy}{dt} = 3a + b \tag{A.1}$$

for all values of time. But if $a = a(t)$, such as,

$$a(t) = \sin(kt)$$

then A.1 becomes

$$\frac{dy}{dt} = \sin(kt)x + b$$

and the value of (dy/dt) does change with time.

Partial Derivatives

Standard introductory differential calculus assumes a function of a single variable, $f = f(x)$, so x is an independent variable and f is a dependent variable. (Often x is time.) The ordinary derivative is defined by

$$\frac{df}{dx} = \lim_{\Delta x \to 0} \frac{f(x + \Delta x) - f(x)}{\Delta x}$$

The derivative is the slope of the function at a particular value of x.

The general case is a function of more than one variable. In a two-variable equation, $f(x, y)$, there are two possible slopes, one for each direction along the plane defined by x and y.

Analogous to the ordinary derivative definition, two partial derivatives of the function, denoted by $\partial f/\partial x$ and $\partial f/\partial y$, are given by the formulas

$$\frac{\partial f}{\partial x} = \lim_{\Delta x \to 0} \frac{f(x + \Delta x, y) - f(x, y)}{\Delta x}$$

$$\frac{\partial f}{\partial y} = \lim_{\Delta y \to 0} \frac{f(y + \Delta y, x) - f(x, y)}{\Delta y}$$

In the $(\partial f/\partial x)$ definition y is held constant and only x varies; likewise, in the $(\partial f/\partial x)$ definition x is held constant while y varies.

The partial derivative of a multivariable function is a direct extension of the definition. So for $f = f(x, y, z, \cdots)$

$$\frac{\partial f}{\partial x} = \lim_{\Delta x \to 0} \frac{f(x + \Delta x, y, z) - f(x, y, z)}{\Delta x}$$

$$\frac{\partial f}{\partial y} = \lim_{\Delta y \to 0} \frac{f(y + \Delta y, x, z) - f(x, y, z)}{\Delta y}$$

$$\frac{\partial f}{\partial z} = \lim_{\Delta z \to 0} \frac{f(z + \Delta z, x, y) - f(x, y, z)}{\Delta z}$$

$$\vdots$$

In each case only a single variable varies and all others are held constant.

In ordinary differential equations, routine mechanical procedures can be obtained from the basic definition of the derivative, such as the familiar

$$f(x) = x^n \quad \text{and}$$

$$\frac{df(x)}{dx} = nx^{n-1}$$

$$\frac{d^2f(x)}{dx^2} = (n)(n-1)x^{n-2}$$

$$\vdots$$

$$\frac{d^k f(x)}{dx^k} = (n)(n-1)\cdots(n-k+1)x^{n-k}$$

These same rules apply to partial derivatives. Given

$$f(x, y) = x^n y^m \tag{A.2}$$

$$\frac{\partial f(x, y)}{\partial x} = nx^{n-1} y^m \tag{A.3}$$

$$\frac{\partial^2 f(x, y)}{\partial x^2} = (n)(n-1)x^{n-2} y^m \quad \text{and} \tag{A.4}$$

$$\frac{\partial f(x, y)}{\partial y} = mx^n y^{m-1} \tag{A.5}$$

$$\frac{\partial^2 f(x, y)}{\partial y^2} = (m)(m-1)x^n y^{m-2} \tag{A.6}$$

and so on for higher-order partial derivatives. Clearly the partial derivative operator is distributable, so if instead of A.2 we had

$$f(x, y) = x^n y^m + x^p y^q \tag{A.7}$$

then

$$\frac{\partial f}{\partial x} = \frac{\partial (x^n y^m)}{\partial x} + \frac{\partial (x^p y^q)}{\partial x} \tag{A.8}$$

which evaluates to

$$\frac{\partial f}{\partial x} = nx^{n-1} y^m + px^{p-1} y^q \tag{A.9}$$

$\partial f / \partial y$ is determined in the same fashion.

In classical dynamical systems theory the variables change over time, t; time is independent and unaffected by the values of the variables; time is an independent variable. With a two-variable dynamical system with x and y time-dependent the function $f(x, y)$ is in fact only a function of time. The overall function f is called the *dependent function* and is really a function of the independent variable t; x and y are *intermediate variables*. One useful shorthand notation to denote this relation is $f = f(x, y; t)$. This means the change in the function f over time t is the ordinary derivative obtained by the algorithm

$$\frac{df}{dt} = \frac{\partial f}{\partial x}\frac{dx}{dt} + \frac{\partial f}{\partial y}\frac{dy}{dt} \tag{A.10}$$

In other words, the total change in the dependent function f with time t is the sum of the change in the two intermediate variables x and y.

Given

$$f(x, y) = 3x^2 + 4xy^3 \tag{A.11}$$

with x and y both functions of time, we can find the ordinary derivative of f with time. By applying the formula of A.7 and A.10 we find

$$\frac{\partial f}{\partial x} = 6x + 4y^3$$

$$\frac{\partial f}{\partial y} = 12xy^2$$

So

$$\frac{df}{dt} = (6x + 4y^3)\frac{dx}{dt} + 12xy^2\frac{dy}{dt} \tag{A.12}$$

If the exact relation of x and y with time is known then both dx/dt and dy/dt could be obtained and A.12 evaluated.

Notice there is no change in this procedure if instead of time, x and y are functions of any other independent variable, c. Then $f = f(x, y; c)$. Also, the procedure is readily extended to three or more variables. So, for $f = f(x, y, z; c)$ with $x = x(c)$, $y = y(c)$, and $z = z(c)$

$$\frac{df}{dt} = \frac{\partial f}{\partial x}\frac{dx}{dt} + \frac{\partial f}{\partial y}\frac{dy}{dt} + \frac{\partial f}{\partial z}\frac{dz}{dt}$$

In ordinary differentiation of a single dependent variable, the *higher-order* derivatives are repeated applications of the differentiation process. With $f = f(x)$

$$\frac{d^2f}{dx^2} = \frac{d}{dx}\left(\frac{df}{dx}\right)$$

$$\frac{d^3f}{dx^3} = \frac{d}{dx}\left(\frac{d^2f}{dx^2}\right) = \frac{d}{dx}\left[\frac{d}{dx}\left(\frac{df}{dx}\right)\right]$$

For a function with more than one intermediate variable the same procedure applies, complicated by more combinations for the higher-order derivatives.

For $f(x, y)$ there are three second-order partial derivatives

$$\frac{\partial^2 f}{\partial x^2} = \frac{\partial}{\partial x}\left(\frac{\partial f}{\partial x}\right)$$

$$\frac{\partial^2 f}{\partial y^2} = \frac{\partial}{\partial y}\left(\frac{\partial f}{\partial y}\right)$$

$$\frac{\partial^2 f}{\partial x\, \partial y} = \frac{\partial}{\partial x}\left(\frac{\partial f}{\partial y}\right)$$

Note: the last term is equivalent to

$$\frac{\partial^2 f}{\partial y\, \partial x} = \frac{\partial}{\partial y}\left(\frac{\partial f}{\partial x}\right)$$

The order of differentiation does not matter for continuous functions.

Differential Equations

In the dynamical systems of this book we represent a rate of change in the *i*th variable x_i over time as a function of its own level (population abundance, in the case of a species), and that of the other variables $x_1, x_2, \cdots, x_n$ in the system. Then

$$\frac{dx_i}{dt} = f_i(x_1, x_2, \cdots, x_n; c_1, c_2, \cdots, c_m, t) \tag{A.13}$$

The c's are independent variables, in fact parameters of the system. For example, for the first equation of a three-variable system

$$\frac{dx_1}{dt} = 3x_1 + 5x_1 x_2^2 + 6x_2 x_3 + 25 \sin t$$

the parameters (independent variables) are $c_1 = 3$, $c_2 = 5$, $c_3 = 6$, and $c_7 = 25 \sin t$; the last parameter is the only one which changes over time. We soon will address the question of what happens when any of the c's change value, that is, there are outside events that are not part of the equations that change the values.

For each differential equation in a set of simultaneous equations, the equilibrium values x_i^* occur when the rate of change is zero, $(dx_i/dt) = f_i = 0$, for all i. (There may be more than one equilibrium value for each variable.) Equation A.13 has the dependent variable, $dx_i/dt = f_i$, and the independent variables c's and t, as well as the intermediate variables $x_1, x_2, x_3, \cdots, x_n$. The equation can be differentiated for each intermediate variable, evaluated at equilibrium

$$\frac{\partial}{\partial x_j} \frac{dx_i}{dt} = \frac{\partial f_i}{\partial x_j}$$

For a linear system this will not matter since there is only one equilibrium solution, $\mathbf{x} = \mathbf{0}$. For example, given

$$\frac{dx_1}{dt} = 3x_1 - 12x_2 + x_3 \qquad (A.14)$$

$$\frac{dx_2}{dt} = -5x_1 + 9x_2$$

$$\frac{dx_3}{dt} = 6x_2 - 10x_3$$

then the equilibrium values of A.14 occur when

$$\frac{dx_1}{dt} = \frac{dx_2}{dt} = \frac{dx_3}{dt} = 0$$

and are $x_1 = x_2 = x_3 = 0$, the origin. If any or all of the linear equations have some independent term in addition to time, such as any constant, then there is still only one equilibrium value for each variable but the equilibrium point of the system is shifted from the origin.

Taylor Series

Nonlinear systems can have more than one equilibrium point. The equilibrium at each set of solutions to $dx_i/dt = 0$ is considered. To do so, the equations are *linearized* around one of the equilibrium points, analogous to a Taylor series expansion.

A Taylor series of a single equation or function is a power series expansion of the function, so

$$f(x) = a_0 + a_1(x - x_0) + a_2(x - x_0)^2 + a_3(x - x_0)^3 + \cdots \quad (A.15)$$

or in compact form

$$f(x) = \sum_{n=0}^{\infty} a_n(x - x_0)^n$$

This is a Taylor series expansion around the value x_0, with

$$a_0 = f(0) \quad (A.16)$$

and

$$a_n = \left(\frac{1}{n!}\right) \frac{d^n f}{dx^n}\bigg|_{x_0 = 0} \quad (A.17)$$

Equation A.16 is the original function evaluated at $x = x_0$, which is usually but not necessarily zero; while A.17 is the nth derivative of the function also evaluated at zero. To illustrate how this is applied, suppose we have

$$f(x) = -15x + x^2 \quad (A.18)$$

and we want the Taylor series expansion at $x_0 = 5$. Then

$$a_0 = f(5) \qquad = -50$$

$$a_1 = \left(\frac{1}{1!}\right)\frac{df(x)}{dx}\bigg|_{x_0=5} = \frac{-15 + 2x}{1}\bigg|_{x_0=5} = -5$$

$$a_2 = \left(\frac{1}{2!}\right)\frac{d^2 f(x)}{dx^2}\bigg|_{x_0=5} = \frac{2}{2}\bigg|_{x_0=5} = 1$$

$$a_3 = \left(\frac{1}{3!}\right)\frac{d^3 f(x)}{dx^3}\bigg|_{x_0=5} = \frac{0}{6}\bigg|_{x_0=5} = 0$$

$$a_4 = 0$$

$$\vdots$$

Thus

$$f(x) = -50 - 5(x - 5) + (x - 5)^2 + 0x^3 + 0x^4 + \cdots$$

and simplifying yields

$$f(x) = 25 - x + 2x^2 \quad (A.19)$$

Clearly Equation A.19 is no more linear than is A.18. In general, the procedure is to ignore all terms of order two and higher. In this case the result is

$$f(x) = -50 - 5x$$

The Taylor series was expanded around the arbitrary value of $x_0 = 5$. If instead we evaluated around the point which causes $f(x)$ to vanish, that is, $x = 15$, then $a_0 = 0$. This always will be the case. Thus when finding a Taylor series expansion of a first-order differential equation around an equilibrium, the first term in the expansion, a_0, will be zero.

For completeness, we point out that a Taylor series is not restricted to functions of a single variable. A function of two variables has the Taylor series expansion around x_0, y_0, as

$$f(x, y) = A_0 + A_1(x - x_0) + B_1(y - y_0) + A_2(x - x_0)^2 \\ + B_2(y - y_0)^2 + 2C_1(x - x_0)(y - y_0) + \cdots$$

where

$$A_0 = f(x_0, y_0)$$

$$A_1 = \left(\frac{1}{1!}\right) \left.\frac{\partial f(x, y)}{\partial x}\right|_{x=x_0, y=y_0}$$

$$B_1 = \left(\frac{1}{1!}\right) \left.\frac{\partial f(x, y)}{\partial y}\right|_{x=x_0, y=y_0}$$

$$A_2 = \left(\frac{1}{2!}\right) \left.\frac{\partial^2 f(x, y)}{\partial x^2}\right|_{x=x_0, y=y_0}$$

$$B_2 = \left(\frac{1}{2!}\right) \left.\frac{\partial^2 f(x, y)}{\partial y^2}\right|_{x=x_0, y=y_0}$$

$$C_2 = \left(\frac{1}{2!}\right) \left.\frac{\partial^2 f(x, y)}{\partial x \, \partial y}\right|_{x=x_0, y=y_0}$$

$$\vdots$$

Jacobian Matrix

Return now to the differential Equation A.13, in which we assume the function is nonlinear. To linearize use a Taylor series expansion for each equation (dx_i/dt, for all i), expanded around one of the vectors of equilibrium values $\mathbf{x}^*$, and ignore second- and higher-order terms. For the first equation

$$\frac{dx_1}{dt} = f_1(x_1, x_2, \cdots, x_n; c_1, c_2, \cdots, c_m, t) \quad (A.20)$$

we can expand around the equilibrium values $x_1^*, x_2^*, \cdots, x_n^*$, or $\mathbf{x}^*$ for short. This is a Taylor series expansion of a multivariable function, so the partial derivative is employed. Also, since the expansion is around one of the equilibrium points, the first term a_0 is zero. To produce linear equations we drop all second- and higher-order derivatives, thus A.20 becomes

$$f_1(x_1, x_2, \cdots, x_n; c_1, c_2, \cdots, c_m, t)$$
$$= \left.\frac{\partial f_1}{\partial x_1}\right|_{\mathbf{x}=\mathbf{x}^*} (x - x_1) + \left.\frac{\partial f_1}{\partial x_2}\right|_{\mathbf{x}=\mathbf{x}^*} (x - x_2) + \left.\frac{\partial f_1}{\partial x_3}\right|_{\mathbf{x}=\mathbf{x}^*} (x - x_3)$$
$$+ \cdots + \left.\frac{\partial f_1}{\partial x_n}\right|_{\mathbf{x}=\mathbf{x}^*} (x - x_n)$$

For the second equation

$$\frac{dx_2}{dt} = f_2(x_1, x_2, \cdots, x_n; c_1, c_2, \cdots, c_m, t)$$

the expansion is

$$f_2(x_1, x_2, \cdots, x_n; c_1, c_2, \cdots, c_m, t)$$
$$= \left.\frac{\partial f_2}{\partial x_1}\right|_{\mathbf{x}=\mathbf{x}^*} (x - x_1) + \left.\frac{\partial f_2}{\partial x_2}\right|_{\mathbf{x}=\mathbf{x}^*} (x - x_2)$$
$$+ \cdots + \left.\frac{\partial f_2}{\partial x_n}\right|_{\mathbf{x}=\mathbf{x}^*} (x - x_n)$$

Continue in like fashion for all the functions. The *Jacobian matrix* is the matrix of first partials of each term of the Taylor series in the expansion around equilibrium

$$\mathbf{J} = \begin{bmatrix} \frac{\partial f_1}{\partial x_1} & \frac{\partial f_1}{\partial x_2} & \cdots & \frac{\partial f_1}{\partial x_n} \\ \frac{\partial f_2}{\partial x_1} & \cdot & \cdots & \cdot \\ \cdot & \cdot & \cdots & \cdot \\ \cdot & \cdot & \cdots & \cdot \\ \cdot & \cdot & \cdots & \cdot \\ \frac{\partial f_n}{\partial x_1} & \frac{\partial f_n}{\partial x_2} & \cdots & \frac{\partial f_n}{\partial x_n} \end{bmatrix}_{\mathbf{x}=\mathbf{x}^*} \quad (A.21)$$

and
$$\frac{d\mathbf{x}}{dt} = \mathbf{J}f(\mathbf{x} - \mathbf{x}^*) \tag{A.22}$$

The results of A.22 are a set of linear, homogeneous differential equations, conveniently written as

$$\frac{d\mathbf{x}}{dt} = \begin{bmatrix} a_{11} & a_{12} & a_{13} & \cdots & a_{1n} \\ a_{21} & a_{22} & \cdot & \cdots & a_{2n} \\ a_{31} & \cdot & \cdot & \cdots & \cdot \\ \cdot & \cdot & \cdot & \cdots & \cdot \\ \cdot & \cdot & \cdot & \cdots & \cdot \\ a_{n1} & \cdot & \cdot & \cdots & a_{nn} \end{bmatrix} \begin{bmatrix} x_1 \\ x_2 \\ x_3 \\ \cdot \\ \cdot \\ x_n \end{bmatrix} \tag{A.23}$$

The elements of A.23, the a_{ij} of the matrix, are the coefficients used in the loop model.

Changing Parameters—The Loop Formula

In the equation

$$\frac{dx_i}{dt} = f_i(x_1, x_2, \cdots, x_n; c_1, c_2, \cdots, c_m, t) \tag{A.24}$$

we assume each x_i an intermediate variable and the c's all independent. Suppose there is a change in one of the c values—c_3 increases, say. This alters the f's of A.24 that contain the c_3. The way each of these functions changes will determine the new equilibrium values of all the x_i's, even for x's whose functions do not contain the c_3 if they are coupled to equations that do. By way of example, consider

$$\frac{dx_1}{dt} = -4x_1 - 2x_2 - 8x_3 \tag{A.25}$$

$$\frac{dx_2}{dt} = -3x_2 - x_3$$

$$\frac{dx_3}{dt} = 136 + 2x_1 + 6x_2$$

where obviously x_1, x_2, and x_3 are the implicit variables. In this example none of the functions shares parameters; there are eight independent parameters:

$$c_1 = -4, c_2 = -2, c_3 = -8, c_4 = -3, c_5 = -1, c_6 = 136, c_7 = 2,$$
$$\text{and } c_8 = 6$$

The equilibrium values of **x** are:

$$x_1^* = 14, \ x_2^* = 8, \ x_3^* = -24$$

If c_6 is reduced by half to a new value of $\tilde{c}_6 = 68$, then not only does the equilibrium of x_3 change because c_6 appears in the equation of f_3 but also the equilibrium values of x_1 and x_2 change. The new values are

$$x_1^* = 7, \ x_2^* = 4, \ x_3^* = -12$$

The problem can be posed in a general way. What happens to the equilibrium levels of each function, f_i, in a system of differential equations when one of the parameters, c_h, changes? The solution is to examine the partial derivative of the function due to the change in the parameter $\partial f_i / \partial c_h$ at or near the equilibrium level.

From the rules of partial derivatives

$$\frac{\partial}{\partial c_h}\left(\frac{dx_i}{dt}\right) = \frac{\partial}{\partial c_h} f_i(x_1, x_2, \cdots, x_n; c_1, c_2, \cdots, c_m, t) = 0$$

with c_h as an independent variable. Then

$$\frac{\partial f_i}{\partial c_h} + \frac{\partial f_i}{\partial x_j}\frac{\partial x_j}{\partial c_h} = 0 \tag{A.26}$$

This will be true for each function, even if c_h is missing from it—as seen in the previous example with c_6 only appearing explicitly in the equation of x_3. Writing A.26 out in detail for a system of n variables

$$\frac{\partial f_1}{\partial c_h} + \frac{\partial f_1}{\partial x_1}\frac{\partial x_1}{\partial c_h} + \frac{\partial f_1}{\partial x_2}\frac{\partial x_2}{\partial c_h} + \cdots + \frac{\partial f_1}{\partial c_h}\frac{\partial x_n}{\partial c_h} = 0$$

$$\frac{\partial f_2}{\partial c_h} + \frac{\partial f_2}{\partial x_1}\frac{\partial x_1}{\partial c_h} + \quad \cdots \quad + \cdots + \quad \cdots \quad = 0$$

$$\vdots \qquad \qquad \vdots \qquad \qquad \cdot + \qquad \cdot$$

$$\vdots \qquad \qquad \vdots \qquad \qquad \cdot + \qquad \cdot \tag{A.27}$$

$$\vdots \qquad \qquad \vdots \qquad \qquad \cdot + \qquad \cdot$$

$$\frac{\partial f_n}{\partial c_h} + \frac{\partial f_n}{\partial x_1}\frac{\partial x_1}{\partial c_h} + \quad \cdots \quad + \cdots + \frac{\partial f_n}{\partial c_h}\frac{\partial x_n}{\partial c_h} = 0$$

Cramer's rule can be applied to A.26 to solve for $\partial x_i/\partial c_h$ with $\partial f_i/\partial c_h$ the known quantity. The term $\partial f_i/\partial x_j$ is exactly the element of the Jacobian matrix of Equation A.21, that is, the a_{ij} elements of the matrix A.23. Consequently, A.26 can be rewritten as

$$a_{11}\frac{\partial x_1}{\partial c_h} + a_{21}\frac{\partial x_2}{\partial c_h} + \cdots a_{1n}\frac{\partial x_n}{\partial c_h} = -\frac{\partial f_n}{\partial c_h}$$

$$a_{21}\frac{\partial x_1}{\partial c_h} + a_{22}\frac{\partial x_2}{\partial c_h} + \cdots \qquad\qquad = -\frac{\partial f_2}{\partial c_h} \qquad\qquad \text{(A.28)}$$

$$\vdots \qquad\qquad \vdots \qquad\qquad = \qquad \vdots$$

$$a_{n1}\frac{\partial x_1}{\partial c_h} + \cdots + \cdots a_{nn}\frac{\partial x_n}{\partial c_h} = -\frac{\partial f_n}{\partial c_h}$$

The solution is

$$\frac{\partial x_i}{\partial c_h} = \frac{\begin{vmatrix} a_{11} & a_{12} & a_{13} & \cdots & a_{1,i-1} - \frac{\partial f_1}{\partial c_h} & a_{1,i+1} & \cdots & a_{1n} \\ a_{21} & a_{22} & a_{23} & \cdots & a_{2,i-2} - \frac{\partial f_2}{\partial c_h} & a_{2,i+2} & \cdots & a_{2n} \\ a_{31} & a_{32} & a_{33} & \cdots & a_{3,i-3} - \frac{\partial f_3}{\partial c_h} & a_{3,i+3} & \cdots & a_{3n} \\ \cdot & \cdot & \cdot & \cdots & \cdot & \cdot & \cdots & \cdot \\ \cdot & \cdot & \cdot & \cdots & \cdot & \cdot & \cdots & \cdot \\ \cdot & \cdot & \cdot & \cdots & \cdot & \cdot & \cdots & \cdot \\ a_{n1} & a_{n2} & a_{n3} & \cdots & a_{n,i-1} - \frac{\partial f_n}{\partial c_h} & a_{n,i+1} & \cdots & a_{nn} \end{vmatrix}}{\begin{vmatrix} a_{11} & a_{12} & a_{13} & \cdots & a_{1n} \\ a_{21} & a_{22} & a_{23} & \cdots & a_{2n} \\ a_{31} & a_{32} & a_{33} & \cdots & a_{3n} \\ \cdot & \cdot & \cdot & \cdots & \cdot \\ \cdot & \cdot & \cdot & \cdots & \cdot \\ \cdot & \cdot & \cdot & \cdots & \cdot \\ a_{n1} & a_{n2} & a_{n3} & \cdots & a_{nn} \end{vmatrix}}$$

In words, the solution for the change in the ith variable due to a change in parameter c_h, is found by substituting the ith column of the determinant of the system with the vector $-\partial \mathbf{f}/\partial c_h$ and dividing this by the original systems determinant. This can be expressed in the loop analysis form as,

$$\frac{\partial x_i}{\partial c_h} = \sum_{j,k} \frac{\partial f_j}{\partial c_h} p_{ij}^{(k)} \frac{F_{n-k}^{(\text{comp})}}{F_n}$$

References

Briand, F., and E. McCauley. 1978. Cybernetic mechanisms in lake plankton systems: How to control undesirable algae. *Nature* 273:228-230.

Caswell, H. 1982. Stable-population structure and reproductive value for populations with complex life-cycles. *Ecology* 63:1223-1231.

Cushing, D. H. 1975. *Marine Ecology and Fisheries.* Cambridge: Cambridge University Press.

de Bach, P. 1974. *Biological Control by Natural Enemies.* Cambridge: Cambridge University Press.

Finerty, J. P. 1980. *The Population Ecology of Cycles in Small Mammals.* New Haven, Connecticut: Yale University Press.

Frogner, K. 1975. Competition in a Container Breeding Mosquito Community: A Field Study and Model. Ph.D. Dissertation, University of Chicago.

Gantmacher, S. R. 1960. *The Theory of Matrices,* vol. 3. New York, Chelsea Publishing Company.

Hildebrand, F. B. 1962. *Advanced Calculus for Applications.* Englewood Cliffs, New Jersey: Prentice-Hall, Inc.

Holling, C. S. 1959. The components of predation as revealed by a study of small-mammal predation of the European pine sawfly. *The Canadian Entomologist* 91:234-261.

——— 1978. *Adaptive Environmental Assessment.* New York: IIASA/Wiley.

Hutchinson, G. E. 1978. *An Introduction to Population Ecology.* New Haven, Connecticut: Yale University Press.

Kelley, K. M. 1982. The Nantucket Bay Scallop Fishery: The Resource and Its Management. Nantucket, Massachusetts: Shellfish and Marine Department, p. 108.

Kerner, E. H. 1957. A statistical mechanics of interacting biological species. *Bulletin of Mathematical Biophysics* 19:121-145.

Keyfitz, N. 1968. *Introduction to the Mathematics of Population.* Reading, Massachusetts: Addison-Wesley.

Leigh, E. G., Jr. 1975. Population fluctuations, community stability, and environmental variability. In *Ecology and Evolution of Communities,* Cody, M. and J. M. Diamond, eds. Cambridge: Harvard University Press, pp. 51-73.

Levins, R. 1966. The strategy of model building in population biology. *American Scientist* 54:421-431.

——— 1974. Qualitative analysis of partially specified systems. *Annals of the New York Academy of Sciences* 231:123-138.

——— 1975. Evolution in communities near equilibrium. In *Ecology and Evolution of Communities*, Cody, M. and J. M. Diamond, eds. Cambridge: Harvard University Press, pp. 16-50.

Li, H. W., and P. B. Moyle. 1981. Ecological analysis of species introductions into aquatic systems. *Transactions of the American Fisheries Society* 110:772-782.

Lorens, C. S. 1964. *Flowgraphs*. New York: McGraw-Hill.

Lubchenco, J. 1978. Plant species diversity in a marine intertidal community: Importance of herbivore preference and algal competitive abilities. *American Naturalist* 112:23-39.

MacArthur, R. 1972. *Geographical Ecology: Patterns in the Distribution of Species*. New York: Harper & Row.

Mason, S. J. 1953. Feedback theory—some properties of signal flow graphs. *Proceedings of the Institute of Radio Engineers* 41:1144-1156.

Paine, R. T. 1966. Food web complexity and species diversity. *American Naturalist* 100:65-75.

Pederson, J. and C. J. Puccia. 1982. New England tide pool communities: Field studies, models, and predictions. *Malecological Review* 15:148.

Pielou, E. C. 1969. *An Introduction to Mathematical Ecology*. New York: Wiley.

Pollard, J. H. 1973. *Mathematical Models for the Growth of Human Populations*. Cambridge: Cambridge University Press.

Puccia, C. J., and R. Levins. 1983. Statistical patterns of non-equilibrium systems. In *Analysis of Ecological Systems: State-of-the-Art in Ecological Modelling*, Lauenroth, W. K., G. V. Skogerboe, and M. Flug, eds. New York: Elsevier, pp. 141-147.

Puccia, C. J., and J. Pederson. 1983. Models and field data of New England tide pools. In *Analysis of Ecological Systems: State-of-the-Art in Ecological Modelling*, Lauenroth, W. K., G. V. Skogerboe, and M. Flug, eds. New York: Elsevier, pp. 709-718.

Roberts, F. S. 1976. *Discrete Mathematical Models*. Englewood Cliffs, New Jersey: Prentice-Hall, Inc.

Rose, F. C., and M. Gawel. 1979. *Migraine. The Facts*. Oxford: Oxford University Press.

Slingerland, R. 1981. Qualitative stability analysis of geologic systems, with an example from river hydraulic geometry. *Geology* 9:491-493.

Spierings, E. L. H. 1980. *The Pathophysiology of the Migraine Attack*. Brussels: Staflev's Wetenschappelijke Uitgeversmaatschappij B. V. Alphen Aan Den Rijn.

Sze, P. 1982. Distributions of macroalgae in tidepools on the New England coast (USA). *Botanica Marina* 25:269-276.

Watt, K. E. F. 1968. *Ecology and Resource Management*. New York: McGraw-Hill.

Wright, S. 1921. Correlation and causation. *Journal of Agricultural Research* 20:557-585.

Index

Bay scallop model, 131–135
Biofeedback, 139

Carrying capacity, 220–221
Competition factor in Lake Tahoe model, 136
Competition model, 70–74
Complementary subsystems, 37; null, 38, 40, 44; symbols of, 38–39; formula for, 40; feedback in, 41–43, 44, 100; and equilibrium of systems, 43, 44; in population models, 93
Complexity, 1–2; reductionist approach to, 4; in loop models, 79, 120
Continuous systems, 239
Correlation among variables, 53, 95–118; from model predictions, 56; zero, 56–59; number of states possible in, 57–58; null hypothesis for, 58; signs of, 58–59; table of predictions for, 58–59; in complex eigenvectors, 184–185; in population models, 201; in predator-prey models, 207–208; in finite time intervals, 217; and carrying capacity, 220–221
Cosine functions, 181
Covariance: in time averaging, 198–200, 218–220; in population models, 201; in predator-prey models, 203, 207–209; in R,X,S system, 210–211; in nonlinear equations, 226; and parameter change, 228–230

Determinants, 163
Diarrhea-nutrition model, 120–122
Differential equations, 243–244
Discrete systems, 190–193, 239
Discrete-time processes, 223–224
Dynamical systems, 238–240, 242, 243

Ecological models, 224. *See also* Fisheries management model; Bay scallop model; Lake model; Lake Tahoe model; Population models; Predator-prey model; Tide pool model
Eigenvalues: in linear systems, 156–158, 166, 173, 178, 180, 182; real and imaginary, 158, 183–184, 186; in polynomials, 166, 168, 176–177; and return to equilibrium rates, 174; and self-damping variables, 178; largest, 179; and associated eigenvectors, 179–180, 187, 190; of vectors, 184; complex, 184; of isoclines, 189; in discrete systems, 190, 192
Eigenvectors: defined, 175; and return to equilibrium rates, 175–176; determination of, 179; and associated eigenvalues, 179–180, 187, 190; complex conjugate, 182; sign pattern of, determined by loop analysis, 182–184, 186; complex, 184–185; and pictorial descriptions, 186–190; of isoclines, 189; largest, 190; in discrete systems, 192; analysis, 193
Engineering systems, 4, 5, 187
Equilibrium in systems, 150–152; and parameter change, 36, 45; abundance, 40–45, 127; symbols for, 41, 43; and overall feedback, 41, 43, 44; formula for, 41–42, 43–44; and complementary feedback, 41–43, 44; and complementary subsystems, 43, 44; in tables of predictions, 49–50; and turnover rates, 53–55; in competition model, 72; $dy/dt = 0$, 150–152; value of, 153–156, 184, 244; rate of return to, 173–176, 184–186; and loop analysis, 194; in pest model, 222–223; in nonlinear systems, 244–246
Euler's formula, 158, 181
Expectation operator in time averaging, 196–199, 204

Experimental variation, 100–101, 106, 117, 138–139

Feedback in loop models, 75; defined, 17; positive, 17–18, 29–31, 93, 107; negative, 17, 19–24, 28–29, 31, 93, 113; level of, 18; calculation of, 18–23, 45; multiplication of, 26–27; of complementary subsystems, 38, 41–43, 44; and equilibrium in systems, 41, 43, 44; zero, 74, 117, 128; in plant population model, 83; in diarrhea-nutrition model, 120, 122; in migraine model, 124–126; in lake model, 128; in small mammal population model, 130; in bay scallop model, 133–135; in linear systems, 164, 179
Finite time intervals, 195; and correlation among variables, 217; and covariance, 218–220
Fisheries management model, 77–79

Generality in models, 10
Graph structure construction: changes in, 137–139; and self-enhancement, 138; and experimental manipulation, 138–139; from data, 139–149; stepwise, 139–142; by matrix inversion, 139, 142–149; by trial and error, 139, 149; ambiguity in, 145–147

Higher-order moments in nonlinear equations, 226
Hurwitz determinants of polynomials, 167, 169, 170, 172. *See also* Routh-Hurwitz

Intervention, experimental, 137–139
Inverse problem in model building, 119
Isoclines: defined, 186–187; in linear systems, 187–188; eigenvector of, 189; eigenvalue of, 189

Jacobian matrix, 159, 246–248

Lake model, 127–128
Lake Tahoe model, 135–137
Leslie matrix, 190–192
Levins, Richard, 11
Life table analysis, 190
Limit cycles, 156
Linear equations, sustained bounded motion in, 224

Linear systems, 238–239; stability of, 153–156, 158–159, 190; matrix notation for, 156–157; eigenvalues in, 156–158, 166, 173, 178, 180, 182; polynomial equation of, 157; oscillatory behavior in, 158; self-damping in, 159–161; negative input into, 161; signed diagraphs in, 162–163; feedback in, 164, 179; loop analysis of, 171–172; matrix analysis of, 172; isoclines in, 187–188; time averaging of, 201; equilibrium values in, 244
Links in loop models: defined, 13, 51; self-effect, 13, 15; notation of, 14, 15; and paths/loops, 14–16; alteration of, 33–34, 111–113; from equations, 51–52; formula for, 51–52; function notation for, 52; removal of, 110; postulated, 117
Loop analysis, 6–7, 50, 119; stability criteria in, 169; of linear systems, 171–172, 224–225; eigenvector sign pattern determined by, 182–184; and equilibrium in systems, 194; and time averaging, 224–225; algorithm, 225, 228; and parameter changes, 225, 248–251; input term for, 228–229; average values in, 233
Loop models: general formula for, 16; feedback in, 17–18, 107; structural change in, 33–35; environmental input in, 51; applications of, 59–74, 120; zeros in, 74–75, 101–106, 112; resource management in, 77–79; grouping in, by rates of change, 79; one-way connection in, 79; intermediate variables in, 80; replacement in, 80, 81; lumping in, 81; ambiguities in, 84–90; opposite sign combinations in, 84–87; experimental design of, 90–95; verification of, 106, 117; removal of variables from, 106–108; saturation factor in, 107–108; new variables in, 109–110; removal of links from, 110; intervention in, 110; link alteration in, 111–113; self-loop in, 112–113. *See also* Links in loop models
Loops: self-inhibition, 13; and paths/links, 14–16; defined, 15; length of, 15, 19, 22, 24, 26, 129–130, 170; conjunct, 17, 19, 26, 170–171; disjunct, 17, 19, 22, 23, 30, 170–171; self-effect, 75; self, 107, 112–113, 117; self-excitation, 117; and stability, 170

Mason, Samuel, 11

Matrix: inversion, 139, 142–149; notation for linear systems, 156–157; Jacobian, 159, 246–248; adjoint of, 180–182; cofactor, 188–189; signed diagraph representation of, 190; discrete, 190; Leslie, 190–192
Migraine model, 124–127
Model building: errors in, 9, 122; strategies for, 9–10; system description in, 119–120, 127; new link inclusion in, 127
Models: uses of, 2; precision in, 4; measurements in, 6–7; limitation of, 7–10; true, 8. *See also* Loop models; Qualitative models; Quantitative models; *specific names of models*

Nonanticipatory systems, 293
Nonequilibrium systems, 233
Nonlinear equations: higher-order moments in, 226; variance change in, 226–227, 230; expected value operation in, 226–228; inputs into, 231
Nonlinear systems, 233, 238–239; stability of, 154–156; equilibrium in, 244–246

Parameter change in loop models, 32–35, 54–55, 81, 90, 91; defined, 33; formula for, 34–35, 139–140, 248–251; and rate of change of variables, 41; and growth rate, 41–43; negative, 42, 44; positive, 44, 45, 49–50; and equilibrium levels, 53–54; in competition models, 70–73; in physiological model, 76–77; response to, 79; in tide pool model, 95, 96–98; natural variation in, 98; in lake model, 127–128; in bay scallop model, 134; and loop analysis, 225; and variances/covariances, 228–230; and variable values, 230; and unknown inputs, 231
Partial derivatives, 240–243
Paths: defined, 14; and links/loops, 14–16; length of, 15, 36; open, 36–40, 44; value of, 36; alternative, 117
Pest model: traditional explanation of, 59; reduction of food supply factor in, 60; alternative prey factor in, 60; feedback in, 60–61, 66–67; table of predictions for, 61, 65; self-damping in, 62; life history factor in, 62–65; stability analysis of, 65; equilibrium levels in, 222–223
Physiological model, 76–77

Pictorial descriptions, 186–190
Plant population model, 82–84, 90
Polar notation for complex numbers, 181–182
Polynomials: of a linear system, 157; feedback in, 164; roots of, 164, 167; characteristic, 164–165, 167, 173, 176–177; coefficient of, 165–167; eigenvalues in, 166, 168, 176–177; Routh-Hurwitz tabulation for, 167–168; stability of, 168–169
Population models, 8–9; bacteria, 79; invertebrates, 79; physiological states in, 80; R,X,S systems, 80–81, 204–206, 210–212; plant, 82–84, 90; small mammals, 88–90, 129–130; fish, 109–110; species introduction in, 135–137; human, 190–192; sustained bounded motion in, 201; covariance in, 201. *See also* Tide pool model
Precision in models, 10
Predator-prey models, 87, 128; single resource, two consumers, 61, 67, 70, 107, 122–123; two resources, two consumers, 70, 206–210; keystone predator in, 107; pseudopredator in, 109; immigration rate in, 122–123; predator density in, 123–124; covariance pattern in, 203, 207–209; variable death rate in, 209; turnover rate of resources in, 215–216; predator response in, 231; stability in, 231; prey abundance in, 231–233
Public health models. *See* Diarrhea-nutrition model; Migraine model

Qualitative models, 32, 50, 91, 119; vs. quantitative models, 4–6; measurement in, 6, 7; abstraction in, 7; applications of, 10–11; variables in, 123–124
Quantitative models, 6–7; vs. qualitative models, 4–6; abstraction in, 7

R,X,S system, 80–81; sustained bounded motion in, 204–206, 211–212; covariation pattern in, 210–211; time averaging in, 211–212
Rates of change, continuous, 190
Realism in models, 10
Reductionism in models, 5
Routh property for the coefficients of polynomials, 166
Routh-Hurwitz: criteria for stability, 28, 29, 172–173, 183, 190; tabulation for polynomials, 167–168

SBM. *See* Sustained bounded motion
Self-damping variable: in loop models, 13, 33; in feedback calculations, 23, 29; in pest model, 62; in plant population model, 83; in tide pool model, 106, 107, 111–113; in diarrhea-nutrition model, 120–122; in migraine model, 125, 126; in small mammal population model, 129–130; in bay scallop model, 131–132; in linear systems, 159–161; and eigenvalues, 178
Self effect terms. *See* Self-damping variable
Self-loops, 107, 112–113, 117
sgn, 183
Sign: opposite, 84–87; ambiguous, 87
Signed diagrams, 90, 95; as loop models, 10–11, 12, 32, 36, 50; symbols of, 12–13; testing of, 58; and interventions, 111–113; qualitative, 119; ambiguity in, 147; in linear systems, 162–163; determinant of, 163; of matrices, 190; in discrete systems, 191–192
Simplification in models, 2, 9, 79
Simulation techniques, 4
Sine functions, 181
Specialization, 2
Stability of systems, 24–31, 32, 84, 86–87, 90, 152–159; local, 24, 27; formula for, 24–26, 28–29; and negative feedback, 28; Routh-Hurwitz criteria for, 28, 29, 172–173, 183, 190; in pest model, 65; alteration of, 79; in diarrhea-nutrition model, 120; in lake model, 128; in small mammal population model, 129; in Lake Tahoe model, 136–137; of linear systems, 153–156, 158–159, 190; and system trajectory, 154; of nonlinear systems, 154–156; of polynomials, 168–169; and loop length, 170
Statistical analysis, 119
Steinmetz notation for complex numbers, 181–182
Stream flow model, 67–70
Summation sign, 237–238
Sustained bounded motion (SBM), 194; calculation of, 200; in population growth models, 201; in biochemical processes, 202–203; in R,X,S system, 204–206, 211–212; in discrete time processes, 223–224; in linear equations, 224; nonlinear, 228
Systems. *See* Discrete systems; Equilibrium in systems; Linear systems; Nonlinear systems; Stability of systems

Tables of predictions: formula for, 45; and variable parameter change, 45, 49–50; construction of, 46–50; and equilibrium levels, 49–50; and turnover rate, 55–56; for correlation among variables, 58–59; in population models, 93; in tide pool model, 96; qualitative analysis, 119; in lake model, 127; ambiguity in, 145–148
Taylor series, 227, 244–246, 247
Tide pool model, 91–98; competition factor in, 96; ephemeral algae, parameter change in, 96–100; correlation pattern in, 96–101; data collection in, 98; experimental manipulation in, 100–101, 106; satellite variables in, 101–106; zeros in, 101–106, 112; self-damping variable in, 106, 107, 111–113; removal of variable from, 106–108; feedback in, 107, 113; saturation factor in, 107–108; new variables in, 109–110; removal of links from, 110; intervention in, 110, 113; link alteration in, 111–113; self-loop in, 112–113
Time averaging methods, 58; defined, 194; over finite intervals, 195, 216–223; expectation operator in, 196–199; variance and covariance in, 198–200, 218–220; differential equations for, 199–216; in R,X,S system, 211–212; and loop analysis, 224–225
Time invariant systems, 239–240
Turnover rates: and equilibrium levels, 53–55; defined, 54; in competition model, 72–75; in diarrhea-nutrition model, 120; of a resource, 215–216

Variables in loop models: defined, 13; growth rate of, 13; increase/decrease of, 17–18; levels of, 24, 36; quantity of, 24, 85, 90; interaction of, 32; rate of change, 41, 150, 190; direction of change, 50–51, 55; satellite, 74–75, 96, 100; principal, 75; intervening, 76; lumping of, 79–84; intermediate, 80; new, 81–82, 109–110; replacement of, 93; in diarrhea-nutrition model, 120; links between, 124, 139; equilibrium value of, 184; complex trajectories of, 194; value of, 195, 199–200, 225, 230, 233. *See also* Correlation among variables; Self-damping variables
Variance: in time averaging methods, 198–200; in nonlinear equations, 226; change in

nonlinear equations, 226–227, 230, 233; and parameter change, 228–230
Variation. *See* Experimental variation
Vector: sign pattern, 179; eigenvalue of, 184; rate of return to equilibrium of, 184–186; of minimum variance, 193

Wright, Sewall, 11

Zeros in loop model graphs, 74–75